WHAT MAKES YOUR BRAIN HAPPY AND WHY YOU SHOULD DO THE OPPOSITE

WHAT MAKES YOUR BRAIN HAPPY AND WHY YOU SHOULD DO THE OPPOSITE

UPDATED AND REVISED

DAVID DiSALVO

FOREWORD BY WRAY HERBERT

Prometheus Books

59 John Glenn Drive
Amherst, New York 14228

Published 2018 by Prometheus Books

Cover image © Exactostock/SuperStock
Cover design by Nicole Sommer-Lecht
Cover design © Prometheus Books

Inquiries should be addressed to
Prometheus Books
59 John Glenn Drive
Amherst, New York 14228
VOICE: 716–691–0133 • FAX: 716–691–0137
WWW.PROMETHEUSBOOKS.COM

22 21 20 19 18   5 4 3 2 1

Library of Congress Cataloging-in-Publication Data

Names: DiSalvo, David, 1970- author.
Title: What makes your brain happy and why you should do the opposite / by David DiSalvo.
Description: Revised edition. | Amherst, New York : Prometheus Books, 2018. |
    Includes bibliographical references and index.
Identifiers: LCCN 2017042461 (print) | LCCN 2017050441 (ebook) |
    ISBN 9781633883505(ebook) | ISBN 9781633883499 (pbk.)
Subjects: LCSH: Happiness. | Logic. | Desire. | Neurosciences.
Classification: LCC BF575.H27 (ebook) | LCC BF575.H27 D57 2 018 (print) |
    DDC 152.4/2—dc23
LC record available at https://lccn.loc.gov/2017042461

To Devin, Collin, and Kayla

CONTENTS

CONTENTS

PART 4: SOCIAL EBBS AND INFLUENTIAL FLOWS

PART 5: MEMORY AND MODELING

PART 6: NOTHING SO PURE AS ACTION

FOREWORD

by Wray Herbert

Early on in this engaging volume, David DiSalvo tells the story of an attempted drugstore robbery that goes badly. It could have been worse—nobody is killed—but two of DiSalvo's coworkers are seriously injured when they take risks way out of proportion to the meager carton of cigarettes that the thieves are after. The drugstore workers aren't reckless people, and they have all of the information they need to make more sensible—and less perilous—decisions. Yet they don't. Instead of calculating their personal risk in a rational way, they make rash and senseless choices. Why?

DiSalvo deconstructs the irrational thinking of these two hapless employees, both of whose judgments were skewed by very common cognitive biases. One young man was blinded by an archaic but potent sexist caricature of womanly weakness—a bias that often acts below the level of conscious awareness. The other worker was also blinded, in his case by a well-rehearsed internal script in which he played the hero. This powerful narrative narrowed his view of the situation's many possibilities, some of them far from heroic. Both acted rapidly and automatically, without deliberation, making choices they would later regret.

Stereotypes and scripts are both valuable cognitive tools, rules of thumb that we all use every day to navigate the world. Without access to these tools, and many others like them, we would be indecisive, paralyzed. But these tools are also potential traps if used inappropriately or applied in the wrong situation—as DiSalvo's coworkers did in trying to foil this attempted robbery. We all trade in stereotypes and scripts all of the time, using them as efficient shortcuts to conserve time and mental energy. But as this anecdote illustrates, they can skew our judgment.

I like this story—and many of the stories that the author narrates throughout the book—because of its ordinariness. The villains aren't clever or ruthless, and the well-intentioned workers aren't unusually foolish. But they are foolish in the way that we all are, in our normal everyday decisions and choices. That's why it's so easy to sympathize with their misguided and earnest judgments and actions.

I also like this tale because it makes two very important points about human behavior and about the way we think and write about psychology:

The first is that human psychology is complicated. I have spent much of my professional life trying to make this point, and it's not as obvious as one might think. The challenge of writing about behavioral science—as opposed to astronomy or immunology or particle physics—is that readers already understand psychology at a very basic level. Everyone has experienced the full range of human emotions, from terror to delight; everyone has been motivated or unmotivated by turns; everyone has exercised self-control and discipline, and failed to. So readers know the territory, the psychic landscape. The challenge for psychological scientists is to uncover another level of scientific understanding that is not obvious, and the challenge for psychology writers is to convince readers that these findings are novel and worth knowing. DiSalvo has a well-earned reputation in the field for rooting his work in life's ordinary details, yet surprising readers with unexpected insights.

I believe that psychology writing is the most difficult of the science-writing specialties. If this seems counterintuitive, it's only because the so-called hard sciences *seem* more abstract and mathematical. They aren't. Indeed, explaining psychological science in a meaningful way is very tricky and requires a highly sophisticated level of scientific literacy. Human behavior is highly nuanced and unpredictable, all in shades of gray. Journalism, by contrast—including science writing—is often too intellectually tidy and anti-nuance. Explaining the subtleties of human thought and emotion and action with the blunt tools of journalism is a formidable task.

The second point is that the lives we lead are much more automated than we imagine—or like to concede—and that this is not always a good thing. This is one of the key ideas to emerge from cognitive psychology over the past decade or more—and it's the idea that DiSalvo chooses to explore here. Our thinking is guided—and, too often, misguided—by not just a few irrational biases, but by hundreds of such biases. They interact one with another, amplify each other at times, and counter each other in other circumstances. Findings from dozens of laboratories around the world are converging on the conclusion that the human mind—despite its unique powers of analysis—is also deeply flawed by these many biases. Just as we lean too heavily on caricatures and scripts to guide our behavior, so too are we shaped by our need for certainty and closure, our desire for social connection, and other cognitive forces that DiSalvo illuminates with such clarity here.

These insights into our automatic decision making are relatively new, and they are entwined with the corollary idea that the human mind is a kind of dual processor. We are constantly toggling back and forth between a plodding, analytical style of thought and a rapid, impressionistic style. We as individuals may tend to be more deliberate or more intuitive in our thinking style, but we're all a mix of calculation and intuition. We're all capable of slow deliberation, but we don't always choose to slow down at the right moment. There are times that call for speed and gut-level thinking, others that call for caution. It's all about the fit, and the fact is we're not all that good at fitting our thinking to the problem at hand. In other words, thinking is messy.

But writing about messiness is not easy, because it's all too easy to get seduced by certain fallacies of explanation. When what we're describing is complex and messy, our urge is to simplify the mess by imagining tidy categories, a tendency we call *reductionism*. We tend to analyze human nature one quirk at a time, rather than acknowledging the sloppiness. By keeping his analysis grounded in examples of real lives, DiSalvo avoids the reductionist pitfall.

Psychology writers also face a special challenge these days, and that is to resist the seductiveness of the brain. The insights in this volume come from paradigm-shifting work within the field of cognitive psychology. These advances have been accompanied by similar advances in the field of neuroscience, including the discovery of the brain's remarkable neurochemistry and methods to watch the brain in action. This rich, new science promises important new explanations of human behavior, and it has sparked a proliferation of books and other writing about brain science. As amazing as these insights are, they are still limited in their power to explain human thinking and emotion.

It's a common mistake to think that writing about the brain is more sophisticated—or more scientific—than writing about the mind and behavior. And many writers buy into that fallacy. But, in fact, the opposite is true. Brain anatomy and brain chemistry are rooted in biology— hence their privileged status in science writing—but these inquiries do not explain nearly as much about human nature as they promise. The simple fact is that the brain does not equal behavior, and reports on brain activity do not necessarily illuminate important questions—like why we do things that are not in our best interest. To get at these intriguing and complex questions, one must do the painstaking work of reading experimental psychology.

There is something of a backlash taking place right now against the reductionism and overpromising of neuroscience, with brain scientists themselves pointing out the limitations of the field. This is not to say that probing and scanning the brain is unimportant, and no doubt we will one day find meaningful answers in this approach, but for now the brain is not an explanation of the nuances of human psychology. For that, we still need to study psychology. DiSalvo wisely does not let himself be enticed by the easier—but less insightful—way of looking.

Serious psychology writers face another special challenge—how to rise above all of the incredibly bad psychology writing on the market. The typical psychology section of most bookstores—often called the self-help section—is full of books pontificating on the human condition. Some of the authors have academic credentials and some do not—but that really doesn't seem to matter. All offer prescriptions for living better, but few of these prescriptions are rooted in science—or any kind of rigorous intellectual inquiry.

What Makes Your Brain Happy and Why You Should Do the Opposite is not a self-help book. Instead, it's what the author calls "science-help." What this means is that DiSalvo has done the hard legwork of visiting laboratories and digesting the scientific literature, and he is now reporting back on the best scientific insights available on human thinking. The prescriptions are modest, as they should be, because the applied part of cognitive psychology is still young. The best that a science writer can offer right now—the most responsible course—is to make readers aware of the many and surprising ways that the human mind trips itself up in ordinary ways every day. Talking ourselves out of irrational and dangerous judgments and decisions remains our responsibility, but DiSalvo gives us some new and valuable tools.

PREFACE TO THE NEW EDITION

When this book was first published in 2011, we had recently entered a new era of understanding the brain-behavior connection. The question, "Why do we think as we think and do as we do?" was taking on new meaning, particularly because neuroscience was increasingly offering ways of examining the question that weren't available even a decade prior.

In the few years since, it's difficult to quantify just how much new research has hit the scene that in one way or another touches on these questions, which are always gaining more attention in both scholarly and popular press. As someone who writes science and health articles for popular magazines, I'm shoulder-deep in new research much of the time, and, as a result, I have a decent perspective on the latest understandings emerging from labs around the world. Allowing for variability in research quality, certain trends are clear in the best of these results.

Viewing these trends over time leads to a few conclusions, and one is that the original thesis of this book is more strongly supported now than even when it was first published. The brain is a prediction and pattern-detection machine with a penchant for storytelling that craves certainty, stability, and predictability. Begin with that understanding, and a great deal starts making sense when we ask, "Why do we think as we think and do as we do?" Begin with that understanding, and you'll also begin making sense of yourself.

That was, and continues to be, the main driver for why I started writing about these topics to begin with: making sense of, specifically, *why I think as I think and do as I do*. As I've mentioned in many interviews since the first edition was published, that was my starting point when I started writing about the subjects in these pages, and it's with this same flavor of introspection that I'm happy to see this book going back into the world with a few new content adjustments and research findings. The thesis that the content orbits around feels more relevant and well-supported now than ever before.

And that's heartening to know, not only because it makes the book in your hand worth reading, but because it shows that *we're getting somewhere*—in the really big sense of that statement. The human brain, and by association human thought and behavior, is a tremendously complex thing to understand, but we're getting closer to true understandings that yield clearer answers.

Having said that, what's also true is that we're always uncovering questions that aren't close to being answered, no matter how much we'd like to claim otherwise. One of the perils of popular-media science treatments is jumping to answers that really don't exist. We'd like them to exist. We'd like answers to guide us. We'd love to create how-to systems around these answers for others to follow. But when you break it down, this is little more than either well-intentioned wish fulfillment or, sometimes, manipulation of peoples' need for answers and ways to change their lives. As a journalist, part of the challenge is to know which way is which, and to check myself continuously when approaching questions that are likely to remain open for a good long while, if not indefinitely.

With that, I leave you to read this edition, which I hope has judiciously followed a path that celebrates what we know while acknowledging what we don't, always preserving the space between.

HACKING THE COGNITIVE COMPASS

"What a peculiar privilege has this little agitation of the brain which we call 'thought.'"
—DAVID HUME, *DIALOGUES CONCERNING NATURAL RELIGION*

"There is always an easy solution to every human problem—neat, plausible, and wrong."
—H. L. MENCKEN, *THE DIVINE AFFLATUS*

OUR BRAINS ARE PREDICTION AND PATTERN-DETECTION MACHINES THAT DESIRE STABILITY, CLARITY, AND CONSISTENCY—WHICH IS TERRIFIC, EXCEPT WHEN IT'S NOT

You enter the office on your first day of work. Nervous energy tingles through every limb, and you are as alert as a deer sipping from an alligator pond. This is not your "first" first day of work; you've started other jobs before, and these sensations aren't entirely new to you. But still, this job is new, and you're nearly as anxious about it as you were before your very first job years ago. There is, however, a major difference, though it isn't explicitly clear to you as you stroll down the central hallway of the office suite for the first time. But, step by step, as you start looking into the offices you pass and absorb the surroundings, something begins happening that triggers a nascent thought: *I'm going to be fine.*

Why does this thought break through the electric jitter swarm and announce itself? What is changing as you make your way down the hall and begin ingesting the sights and sounds around you? While it's hardly obvious, your brain is doing some heavy lifting on your behalf. Everything you see, smell, touch, and hear is being processed, analyzed, and decoded. Your brain is doing what it has evolved to do, and it's doing it exceptionally well; so well, in fact, that you are starting to experience an emotional response that counteracts your nervous response. Your brain is determining that you have been here before. Not literally, of course, but your brain is structured to make sense of stimuli and patterns in any environment you step into, and it's finding patterns in this new environment that overlay well with others you have experienced. Your brain is arriving at a determination that these patterns are familiar enough that you will be able to make reliable predictions about what is coming next in this environment. As you begin meeting people in the office, more stimuli are processed, more patterns are detected, more is added to the webs of information your brain creates about everything you experience. The more the day goes on, the more at ease you become about most things in this new environment, and those things that have put you on guard have been flagged as potentially dangerous and requiring elevated attention. During the course of one day, your brain has mapped out a new microworld that you will inhabit for as long as you have this job. It will be added to and subtracted from, shifted, adjusted, and contorted—but all of these movements will occur within a framework derived from recurring patterns that your brain has identified, coded, and categorized.

Years of neuroscience research have led to the current understanding of the brain as a prediction machine—an amazingly complex organ that processes information to determine what's coming next. Specifically, the brain specializes in pattern detection and recognition, anticipation of threats, and narrative (storytelling). The brain lives on a preferred diet of stability, certainty, and consistency, and it perceives unpredictability, uncertainty, and instability as threats to its survival—which is, in effect, our survival.

The problem is that our brains' evolved capacity for avoiding and defending against these threats—a capacity that has allowed our species to survive and thrive—has a slew of by-products, all tightly woven into our day-to-day thinking and behavior. This book will discuss several of them, each of which, ironically, trip and ensnare us while making our threat-anticipating brains "happy." The pages ahead include explorations as to why:

- We crave certainty and the feeling of being right.
- We rely on memory to buttress that feeling.
- We're prone to assigning meaning to coincidence, and making causal links with scant information.
- We want to feel in control.
- We try to avoid loss.
- We regulate our moral behavior to feel "balanced."
- We attempt to circumnavigate regret.
- We generalize when specificity would be more beneficial.

If we could live our lives without bias, distortions, and delusions involved, the world would be truly idyllic. But we can't, though we're largely ignorant of this fact. We function much of the time with an air of mystification about why we do what we do, and why we think as we think—not because we are dull-witted. Much the opposite: only a brain advanced enough to engage in complex thought and self-reflection is susceptible to the fuzzy mystification that obscures from view how our minds really work.

Before we go much farther, though, let's take a couple of steps back and discuss where we have been, cognitively speaking, and where we are going.

HACKING THE MISUNDERSTANDINGS OF MIND

For any analysis of mind to be useful, it must defer to what we know about how our brains function. Admittedly, this knowledge is limited, but it has grown enormously in the last few decades, providing an understanding that few thought possible a century ago. If you had, for example, told an early twentieth-century neurologist that technology would develop in the next one hundred years that allows paraplegics to control robotic arms with their minds, you'd probably get a sneer, if not a snicker. Science fiction novels and comic books featured similar technology, but serious scientists wouldn't have staked their careers on such things being possible. We now know that they are more than possible—they are happening. Likewise, we have learned enough about the brain to know that the mind–body dualism of old is an outmoded explanation. It's still tempting for many to yank the workings of mind from their biological moorings, chiefly because the complexity of thought seems too large an enigma for our brains to contain. As one biology professor at my alma mater put it, "How can billions of on-off switches result in something as complex as the mind?"

Cognitive science has not exhaustively answered those sorts of questions, but in the course of diving into the brain's mysteries cognitive science has discovered that the questions themselves were never really on target. The "on-off switch" analogy, for instance, is the result of a subtle category error. By starting with the belief that the brain is essentially a fleshy and compact electrical device—albeit a complicated one—it's impossible to arrive at an explanation for mind that passes anyone's laugh test.

Cognitive science challenges our categories by breaking down the mental silos we build to make sense of things. Consider the temptation to pinpoint thoughts and emotions in well-defined parts of the brain. It's neater to believe that anger, for instance, launches from one central place than to accept that it doesn't "live" in any single place in the brain but is rather the result of multiple brain regions cross-activating in less-than-tidy ways.

This is an especially hard realization to accept when it comes to memory. Where does the memory of riding a roller coaster at Six Flags when you were ten years old reside? Because our recollection of the event seems more or less complete, we want to believe that it must exist that way on a bookshelf in our heads. When we want to revisit the memory, we pull the book from the shelf and turn to the right page. We now know that memory doesn't work that way, and, in fact, your memory of the hairpin curve and corkscrew loop doesn't really reside in any single place in your brain, nor is it in any way complete.

These understandings are all quite messy, and the science underlying them doesn't satisfy our hunger for airtight answers. We jump back into categories to fill the voids because not having answers is unnerving, and so it should be. Since the very organ that defies explanation evolved to make sense of our environment, it's perfectly understandable that we'd become frustrated by the brain's silence about its own inner workings. And yet, the reality is that you and I can carry on this discussion precisely because the amazing organ in our heads yields this thing we've come to call *mind*. Or we could more accurately say that mind is not something produced by the brain, but that which the brain *does*. Said still another way, the brain's activity—and, indeed, the activity of our nervous system in total—*is our mind*. To quote neuroscientist Simon LeVay, "The mind is just the brain doing its job."[1]

For the better part of a century, we have steadily moved away from the idea that the body (including the brain) and mind are separate entities—a belief popularized by what the seventeenth-century French philosopher René Descartes and labeled the mind–body problem, or "dualism." Where dualism went wrong, to paraphrase the contemporary philosopher John Searle, was "to start counting in the first place."[2] But the reason for the "counting"—the bifurcating of brain and mind—is easy enough to see: for as long as humans have been able to think about it, we haven't liked the alternatives. If mind is what the brain does, then it can be reduced to biological processes. And no matter how complex these processes are, they are still the workings of flesh, blood, cellulose, and sinew. How can we—the magnificent, above-common-nature creatures we believe ourselves to be—be tethered so crudely to nothing more than what some neuroscientists call *wetware* (the biological corollary to computer hardware)? That's the challenge to our self-understanding that cognitive-science research presents us with, and it will only become stouter as more revelations about how the brain works come to light.

HACKING FOR BETTER ANSWERS

With dualism behind us, what's in front of us? The comfort of locating the mind apart from the wetware we carry around in our skulls is gone, so what, exactly, should replace it? The answer is central to the argument I'll be making throughout this book. We have entered a period of self-understanding only vaguely imaginable before the new wave of neuroscience and cognitive-psychology research opened the door and began pushing us through. We are at only the beginning of this period, and caution is warranted about drawing hasty conclusions from a body of research still in its infancy. But we are definitely on a new path to self-understanding, and there is no returning to the backwater refuge of dualism. In this new period, when we speak of mind we are speaking of what our brain does. When we speak of thought, we are speaking of the currency of mind—the very stuff of the brain's relentless activity. The dualistic division, figment though it was, has collapsed, and with it died a thousand misconceptions of mind.

What this all points to is a fantastic opportunity—an opportunity to credibly figure out why we do what we do, and, just as important, decipher how we can alter thought and behavior inconsistent with our best interests. If that statement strikes you as having an air of "self-helpness" about it, let me correct the perception in advance: I believe that the new wave of cognitive research actually undercuts a great deal of self-help advice, and will continue to do so in the years ahead by showing just how vacuous, groundless, and fraudulent a percentage of that advice really is. The fog of misunderstandings about the brain and mind has allowed certain varieties of self-help snake oil to flow with impunity for decades, fueled by billions of dollars from well-intentioned consumers looking for answers. Cognitive science cannot provide a complement of concrete answers to replace those of the self-help industry, nor should the disciplines within psychology attempt to do so. What neuroscience and psychology can do, however, is make us smarter evaluators of our thought and behavior by shedding much-needed light on difficult questions. By using sound research as a basis for rethinking our behavior, we will be on steadier ground than much of the self-help industry could provide. We do not need more self-help—we need more *science-help*.

HACKING WITH HUNCHES

I'm a pragmatist. I have a penchant for what works and tend to critically scrutinize assertions that lean on a hunch. But I also understand and appreciate that sometimes a hunch is all we have, and that although it's not the complete answer, it might eventually guide us to one. Research does not operate outside the world of hunches. The best researchers I have met are world-class hunch makers who sometimes craft the most creative and compelling research approaches from a hunch they had while eating breakfast. From a mere hunch, they sometimes arrive at a new understanding that a volume of previous research on the same topic somehow missed. My tour of such research leading up to this book has demonstrated to me that, at times, it pays to have faith in hunches.

Along with that faith, however, it doubly pays to check one's naïveté more often than one might think necessary. Part of what animates the self-help juggernaut and, more recently, the burgeoning industry built on hasty conclusions about neuroscience research, is a naïve approach to solving problems. We want answers. We want to listen to people who claim to have answers. We want problems solved and settled so that we can feel good about the resolution. It hurts to realize that more often than not, we can't have what we want, or at least not as we envision it. But naïveté is a formidable force with the power to trump healthy skepticism about what we are being told is "the answer." If we are not careful, a sincere desire to figure things out can lead to a naïve acceptance of well-heeled nonsense.

As an example, we must be extremely careful about drawing ironclad conclusions from brain imaging studies. The neuroscience community is far from united about what the activation of various brain regions means in all cases. A deep well of issues must still be addressed before the images can speak to us with clear answers. For instance, why, from one study to the next, do different regions activate under the same testing conditions? That the brain is such an effective foil of study replication is a true problem for researchers, and so far no one has arrived at a foolproof way to solve it. Some have gone so far as to argue that we should use brain imaging in the courtroom as proof of guilt or innocence—a truly frightening prospect for a technology that is not ready for that challenge, and arguably never will be.[3] Many other issues could be mentioned, but suffice it to say that the business of science is not to provide us with settled answers that we can comfortably rest our heads upon at night. Indeed, we are wise to expect more new questions than answers from any research campaign worth discussing.

Having said that, the very process of scientific investigation—one finding building on the next with confirmation or challenge—is hope-inspiring. What differentiates scientific assertions from the droves of poorly grounded self-help and pseudoscience assertions is this process. It demands far more of its executors because the process is, in a sense, bent on self-destruction. It doesn't trumpet the perfection of its outcomes; it calls out for challenges that could very well undermine the outcomes and start the process anew.

That, in short, is where this book begins. Science is a tool—but it is the best tool we have to address hard issues about ourselves and our world. And I believe it is also the finest tool available to understand what is catalyzing our thoughts and motivating our behavior. If we are to credibly claim knowledge of why we think as we think and do as we do, then we must engage these questions at their core, and accept the limitations inherent in this process of discovery.

HACKING AHEAD

A few upfront disclosures are in order. First, throughout this book I will use an intentionally oversimplified metaphor of a "happy brain." Brains, of course, cannot strictly speaking be happy or sad or angry, nor can they want, desire, claim, or commit. In the words of New York City–based clinical psychologist and psychoanalyst Todd Essig:

> Brains don't want, any more than lungs sing or knees set long-jump records. Brains are part of what makes people want and how we want. There is always a situated, contextualized, enculturated person between the brain and wanting.[4]

What I wish to communicate with the metaphor of a happy brain is simply that under various conditions, our brains will tend toward a default position that places greatest value on avoiding loss, lessening risk, and averting harm. Our brains have evolved to do exactly that, and much of the time we can be thankful they did. However, these same protective tendencies (what I am calling the tendencies of a "happy brain") can go too far and become obstacles instead of virtues. Our challenge is to know when to think and act contrary to our brain's native leanings.

Second, this is not a book about psychological pathologies. I am not a psychologist or psychiatrist and have no interest in playing virtual therapist via a book or any other medium. I am also not a neuroscientist and would not claim to possess a grasp of neural dynamics that only a full investment in the discipline can provide. I am a science writer especially interested in how our brains work, and I am driven by a passion to communicate what I learn to a broader audience. I am also a public-education specialist who has spent years devising and implementing strategies to boost awareness and catalyze behavioral change among particular target audiences—some narrow, some massive. I am closely familiar with the gap between knowing and applying. Most of us can grasp the substance of a problem and even be provided with a means for overcoming it, yet we often still fail. It is this gap between awareness and action that set me on a path to write this book. I wanted to know why humans so often do things not in our best interest. More specifically, I wanted to understand what attributes of our brains underlie the self-undermining thoughts and actions that plague every person born on this planet.

When I started this trek a few years ago, I expected to focus mainly on cognitive bias—the well-documented throng of mental errors that so often cause us to stumble.[5] But after working through reams of research studies and discussions with experts in the fields of cognitive psychology and neuroscience, I discovered an even more essential piece of the cognitive puzzle, and it has everything to do with what makes our brains "happy."

My investigation also led to a further conclusion: Simply knowing how our brains flop is not very useful. Most books about brain errors never get beyond this point. But what good is *knowing*, if we fall short of *doing* anything about it? We may *know* that we should take action to avoid temptation (for example), but applying that knowledge is a different matter entirely—and that, too, is part of our neural reality. This is the "gap" between awareness and action, and as a practical matter it's just as crucial as figuring out what makes our brains tick.

Finally, what you will find in the following chapters is a broad survey of topics. I have intentionally not dropped too low into the weeds with technical minutia, but rather have focused on what I believe are the larger issues relevant to the discussion. My goal is for this book to be informative but also useful. I hope you will find it to be both as we continue our discussion in the pages ahead.

Part 1

CERTAINTY AND THE SEDUCTION OF CHANCE

Chapter 1

ADVENTURES IN CERTAINTY

"Doubt is not an agreeable condition, but certainty is an absurd one."

—VOLTAIRE, FROM A LETTER
TO FREDERICK II OF PRUSSIA

MIND FULL OF SHARKS

On October 9, 1997, observers from the Point Reyes Bird Observatory witnessed a killer whale clashing with a great white shark near Farallon Island, twenty-six miles off the coast of San Francisco. The sight made for salacious nature news. Speculation about what would happen if these apex predators met has always piqued curiosity, but until that day no one really knew for sure. Someone on the ship caught the confrontation on video, which later made its way onto the Internet and became an instant draw for millions of eyeballs worldwide.[1]

Turns out, it wasn't much of a fight. The orca had little trouble dispatching her menacing opponent, and then proceeded to dine on its liver, leaving the carcass for seagulls to pick clean. This outcome may have disappointed many who expected a bloody, jaw-to-jaw battle between these titans of the deep, but it tickled the fancy of academics to the point of giddiness.

The reason for their interest had to do with why the two clashed in the first place and exactly how the orca defeated the shark. Ordinarily, apex predators are happy to avoid each other, for the simple reason that fighting a beast in your weight and ferocity class will probably result in injury. Injury means impaired ability to hunt, and that means *game over*.

Knowing this, scientists were eager to know why two of the most successful predators on the planet would risk confrontation in the open seas. The answer shocked everyone. This was no chance street brawl: The orca was actually hunting the shark.

To understand why, we have to take a step back to examine how killer whales learn their namesake trade. Like humans, orcas have culture. But unlike most human cultures, orca cultures revolve around one thing: hunting behavior. Some orcas hunt herring, others seal, others stingrays, and others—sharks. The observers on the ship had witnessed an orca conducting the business of its shark-hunting culture.

The next discovery was how the orca so handily defeated the shark. In every orca culture, a hunting technique is learned through demonstration and imitation. That's a big part of what makes orcas such efficient predators—they learn the best, tried-and-true hunting techniques from each other. When one orca tries a killing method that works well, others take notice and copy it.

Scientists speculate that at some point, an orca discovered that if it rammed a shark hard enough from the side, the shark would flip over and become motionless, unable to defend itself and inflict injury. In effect, that pioneering orca induced "tonic immobility" in its adversary—a temporary state of paralysis many species of sharks fall into when turned on their backs. The human discovery of tonic immobility in sharks is relatively recent, making the orca's behavior all the more remarkable.[2]

This deadly shark-hunting technique, capable of rendering a great white shark powerless, is the orca equivalent of a human "meme"—a unit of cultural ideas and practices transmitted from one mind to another. Susan Blackmore, author of *The Meme Machine*, puts a finer point on it by defining a meme simply as "that which is imitated."[3] The biological corollary to a meme is, of course, a gene, a unit of heredity transmitted from an organism to its offspring. Killer whales are, as a matter of heredity, powerful hunters, but we now know that their cultures strongly influence how they use their native abilities. An orca from a herring-hunting culture is not likely to tackle a great white shark, just as an orca from a whale-hunting culture would have no reason to start hunting stingrays.

The key point is that orca cultures pass along memes that benefit their members via learning and perfecting crucial skills necessary for survival. The orca brain is advanced enough to make this meme transfer effective beyond what any other creature in the ocean is capable of achieving. In other words, just about anything might end up on the menu.

The human brain, in contrast, is the undisputed learning master on the planet. Our cultures are infinitely more complex than orca cultures, because the sheer volume and depth of memes we exchange is orders of magnitude greater. The flip side of this reality is that our big brains, advanced as they are, come with an array of complex shortcomings and are also expert at transmitting these shortcomings.

One of the most perilous gene–meme double whammies that humans possess is the notion of certainty. Our natures and our learned biases lead us to believe that we are right whether or not we really are. This is the orca equivalent of learning the wrong way to hunt a great white shark—not a mistake any smart orca would copy. If orca cultures passed along memes that imperiled their members, they wouldn't be long for this world. Humans, on the other hand, pass on problematic memes like the notion of certainty on a daily basis. Rarely does this go well, but rarely does that stop us.

The reason for our stubbornness goes deeper than we think. Neuroscience research is revealing that the state of not being certain is an extremely uncomfortable place for our brains to live: The greater the uncertainty, the worse the discomfort. A 2005 study conducted by psychologist Ming Hsu and his team found that even a small amount of ambiguity triggers increased activity in the amygdalae—two deep brain structures that play a major role in our response to threats.[4] Each amygdala is a cluster of nerve cells that sits under a corresponding temporal lobe on either side of the brain. Information pours into the amygdalae from multiple sources; the amygdalae filter through the information to determine its threat-level significance and mobilize a response. At the same time, the brain shows less activity in the ventral striatum, a part of the brain involved in our response to rewards (we would expect to see increased activity in the ventral striatum when we are anticipating a pay raise, or vacation, or even a kiss, for instance). As the level of ambiguity increases, amygdalae activity continues to increase, and ventral striatum activity continues to decrease.

A 2016 study on a psychological dynamic referred to as the "backfire effect" further underscores just how tenaciously our brains hold onto a sense of certainty in our beliefs. In this case, researchers used political beliefs as the focal point. Study participants were placed in an fMRI scanner and presented with counterarguments to strongly held beliefs with a political slant, such as "Laws restricting gun ownership should be made more restrictive," and "Gay marriage should not be legalized." The analysis revealed that the same parts of the brain that respond to physical threats are also the parts that respond to belief-based threats—and, as in the earlier study, the amygdalae feature prominently in this threat response.[5]

What this tells us is that the brain doesn't merely prefer certainty over ambiguity—it craves it. Our need to be right is actually a need to "feel" right. Neurologist Robert Burton coined the term *certainty bias* to describe this feeling and how it skews our thinking.[6]

The truth for us all is that when we feel right about a decision or a belief—whether big or small—our brains are happy. Since our brains like being happy, we like feeling right. In our everyday lives, though, feeling right translates into being right (because if we could admit that we only "feel" right, then we might not really be right, and from our brains' point of view that's just not alright).

Our fierce mammalian cousins in the oceans are not strapped with the existential baggage of craving certainty. Their needs are far more straightforward, and their brains evolved to facilitate learning specific to meeting those needs. As one unfortunate great white found out, orca brains are very good at what they do.

Our brains are also very good at what they do, but as a consequence of their expansive abilities, our paths to surviving and thriving are not nearly so clear-cut. Our intense desire for feeling right is but one example of this uniquely human reality, and what this chapter is all about.

BLINDED BY THE BLEEDING OBVIOUS

Meet Phil, a youth program specialist at a school for deaf and blind students, responsible for the well-being and mentoring of students living at the institution. Phil (who, by the way, is quite a smart guy—Mensa member to boot) recalls a situation when he started the job: He was making nightly rounds of all of the floors in the blind student dorm to ensure that every student was in his or her room and accounted for. In his previous experience at other institutions, room checks were synonymous with "lights out," but in this case he was instructed that blind students often sleep with their lights on (because lights, on or off, don't matter to them either way) and the administration preferred that the lights stay on for safety reasons.

As he made his rounds, floor after floor, he found that all of the students' lights were on and in each case a student was in the room. When he came to a room with the lights off (the exception to what was now a well-established rule) he walked into the darkness and called out the student's name from his roster. No answer. He called again, more emphatically. Still no answer. After a third panicky call and no response, he checked all of the remaining rooms, bathrooms, and hallways, and, still not finding the student, rushed to the administration office to report him missing. Phil was asked if he was absolutely sure that the student was missing, and he affirmed that he had thoroughly inspected the entire building and "was certain" that the student was not in the room or anywhere else in the vicinity. His statement triggered a campus-wide search for the young student that spilled out well into the city and went on for hours.

At some point during the search, something occurred to Phil that sent nervous energy tingling through his limbs. He ran back up to the floor of the student's room (still entirely dark), blindly reached around an inside wall, and flicked on the light switch. The student was lying comfortably in his bed with earphones on.

How did Phil overlook something that in hindsight seems so obvious? Let's rewind and see what happened. First, Phil was introduced to a new "rule" for success: when lights are on, success is achieved. In his previous positions, the reverse was true, so his brain recalibrated to the parameters of the new rule. He then experienced multiple instances of the lights being on—room after room, floor after floor. These experiences reinforced his brain's recalibration and solidified the new rule.

To put all of this another way: Phil's attention became exceedingly selective. A change to the rule tripped his attention alarm, and the urgency of the alarm overrode consideration of other options. Phil became blind to details that could have changed the outcome—specifically, turning on the light. Phil's behavior is an example of "selective attention," also called *selectivity bias*—the tendency to orient oneself toward and process information from only one part of our environment to the exclusion of other parts, no matter how obvious those parts may be.

Psychologists have uncovered how this dynamic works by using a research method called the Eriksen Flankers Task.[7] Participants are shown three sets of symbols—a middle symbol flanked by a symbol on either side—flashed briefly on a screen. In some cases the flanking symbols point toward the middle symbol (these are called congruent symbols), and in some cases they point away from it (incongruent), and in some cases neither (neutral). After each symbol set is flashed, participants tell the researchers whether the symbols were congruent, incongruent, or neutral, and are also asked to rate how confident they are in their response.

The results are remarkably consistent: Participants say they are highly confident in their responses, but they end up being wrong more than half the time. The reason is that it's shockingly easy to influence the brain to ignore a large part of its environment. By simply flashing the symbols in a pattern and then changing the pattern, the brain remains selectively focused on one variable to the exclusion of others—it simply does not "see" them. Time is a big part of the flankers task. The symbols are purposely flashed for just a moment, forcing the participants to make a quick determination before the next set is flashed. When more time is allotted between sets, responses significantly improve.

By far, the most entertaining research illustrating how extreme the selectivity effect can be is the "Gorillas in Our Midst" study by psychologists Daniel Simons and Christopher Chabris.[8] Study participants were asked to watch a video of a group of people passing a basketball and count how many times the ball is passed. While they are counting, a woman dressed in a gorilla costume slowly walks into the scene, stops halfway to beat her chest, and then slowly walks out of the scene, for a total of nine seconds on screen. After the video ends, participants were asked to answer a few questions, such as "Did you see anything unusual in the video?" and "Did you notice anyone or anything other than the basketball players?" Finally, they were asked, "Did you see the gorilla?" More than half of the participants replied that they had not seen anything unusual, and certainly not a gorilla.

Simons and Chabris successfully catalyzed selective attention by telling the participants to focus on the ball and count the passes. Following this pattern, most of the observers never saw the bizarre sight that appeared right before their eyes.

Participants in these studies report that they are shocked at just how wrong they were. People who complete the flankers task frequently say they were "certain" they had it right. People who complete the gorilla study are amazed they missed something so obvious.

Coming back to Phil, as long as the rooms he was inspecting were all the same, he could effectively judge them as "right" or "wrong" with very little time. In fact, this part of his job became so easy that he was probably flying through it—getting faster as he went. When he came to a room detached from the pattern, he didn't slow down the judgment train a bit; the result was that he didn't see what was right in front of him (albeit in the dark).

What could Phil have done differently? The answer probably looks obvious by now—he should have *slowed down*. Another few moments of deliberation would likely have opened his cerebral eyes to details he was leaving out. But, to do so, he also would have had to challenge his sense of being right—his marriage to certainty. Just as flankers task participants are shocked that they are wrong, Phil, we can safely assume, was shocked that he'd missed such an obvious detail. At least Phil's story ended relatively well, which is not always the case for those in the certainty jungle, as we'll soon see.

DRUGSTORE COWBOYS

With great injury to my teenage sense of entitlement, I worked at a drugstore in my mid-teens to earn enough money for idle pursuits. One day, I was running the cash register when a man with a subtle but noticeable nervous twitch approached the sales counter. He said he had questions about what type of film to use in his new 35 mm camera and motioned to several different boxes of film displayed behind me.

As I turned to pull down a couple of boxes to show him the difference between 200- and 400-speed film, I noticed a woman wearing a large overcoat milling around the front of an aisle where cartons of cigarettes were stacked (back then, some stores still sold cigs on the shelves). I continued talking to the man but kept an eye on the woman as well. The man noticed that I was distracted and started talking faster to get my attention back on him. A few seconds later, I saw the woman stash a carton of cigarettes in her overcoat. That's when it dawned on me that the man and woman were working together—he was distracting me while she looted the joint.

I grabbed a phone and dialed up Ed, the manager in the back office. At this point, the man and woman realized they were caught and both started speed-walking toward the door, with a "the jig is up but look inconspicuous anyway" shuffle. Ed sprinted to the front of the store to stop them from leaving. Within seconds, he had a choice to make: He could either try to stop the man or try to stop the woman, but he could not stop both. One was sure to get out. His subjective determination was that the woman would be easiest to stop, so he reached out and grabbed her by the shoulder. Bad decision. She grabbed his hand, swung around to face him, and while he looked on in horror, pulled back his index finger until it audibly snapped. He fell to his knees, yelling in agony, and both the man and the woman ran out the door.

Ed's decision was made with simple, bifurcated ingredients: man stronger, woman weaker (predicated on his preexisting belief that, without exception, women are the "weaker" sex). His dedication to that dubious logic led him to act, as it turned out, at significant peril. This, however, is not the end of the story. Shortly after Ed's painful miscue, another employee (we'll call him Ned) took off after the pair. Ned was larger than Ed and often boasted of his martial arts skills. He chased the thieving duo into the parking lot, yelling for them to stop—and stop they did. The man turned and confronted Ned, who immediately assumed a sort of hybrid judo–karate stance as a warning to the thief that he'd be no match for Ned's estimable skills. Regrettably for Ned, the man didn't care about his stance or skills, and punched him solidly under his right eye. Ned fell to the pavement, and the thieves continued their escape, this time all the way to their car and out of the parking lot.

The common theme for Ed and Ned is that they both narrowly "framed" the situations they faced, with limited and skewed information, and without consideration of information that could have changed the outcome. The best way to think of how our brains frame information is to imagine a picture frame, except unlike a normal frame, this one obscures everything on the outside of the frame and magnetically draws attention to the inside. If you attempt to look outside the frame, your brain sounds an alarm that it's uncomfortable and wants your attention back squarely on what's inside the frame.

We can see how this tendency affected Ed and Ned. Why, for example, did Ed physically engage the thieves at all? His internal framing led him to believe that he could overpower one of them—the one he presumed must be weaker. In fact, the strength of his internal framing was greater than whatever amount of apprehension he may have felt about physically taking on a total stranger.

Ned's framing went like this: "I can dominate the thieves with my superior fighting skills." He didn't consider possibilities outside of this frame—namely, that he may not be as skilled as he believed, or that the thieves might be more skilled, or even armed. Nor did he consider information he'd already witnessed: Ed's finger snapping like a pretzel.

Psychologists Amos Tversky and Daniel Kahneman were the first to identify this tendency as a cognitive bias (which they called simply *framing bias*), defining it as the decision maker's conception of acts, outcomes, and contingencies associated with a particular choice. The frame that a decision maker adopts is controlled partly by the formulation of the problem and partly by the norms, habits, and personal characteristics of the decision maker.[9]

Ed and Ned were making "on the spot" decisions, and their internal framing manifested without much deliberation (we can cut them some slack for that). If we could peel back the consciousness of these fellas, we would uncover a deeper level of internalized framing, which might better be called *preframing*.

The philosopher and psychology writer Sam Keen tells a story about how he and his brother were labeled by their parents. Sam's brother was a mechanical genius from an early age. He could take apart and reassemble a lawn mower before he was six. His parents told everyone that he was a born engineer—a mechanical marvel. Sam, on the other hand, was "sensitive and thoughtful." He had no discernable mechanical skills, and his parents were careful to point out—to Sam and to others—just how different he and his brother were. Later in life, Sam was trying to figure out whether to go to college or vocational school, so he took a vocational exam to determine his strengths. Afterward, Sam was asked to discuss his results with a guidance counselor. The counselor informed him that he had scored well across the board, but especially well in mechanical aptitude. Sam, shocked by the results, replied, "Oh, you must have the wrong results. That's my brother."[10]

Keen's story reveals something insidious about framing's connection to the feeling of certainty: It operates in our minds like a script written by other people and colored by influences of which we're not even totally aware.

Framing is, of course, also an external influence that affects us in the present. A great deal of research has been conducted on the role of media in framing, particularly when it comes to couching sensitive issues. Kahneman and Tversky devised a classic study to illustrate how this flavor of framing works, re-created below. Consider each scenario and think about your response.

You work for the Centers for Disease Control and there is an outbreak of a deadly disease called "the Asian Flu" in a town of six hundred people. All six hundred people in the town are expected to die if you do nothing. Someone has come up with two different programs designed to fight the disease:

With Program 1, two hundred people in the town will be saved.

With Program 2, there is a 1:3 probability that six hundred people will be saved and a 2:3 probability that no people will be saved.

Which would you pick?

Then another expert comes to you and presents two additional program options:

With Program 3, four hundred people in the town will die.

With Program 4, there is a 1:3 probability that nobody will die, and a 2:3 probability that six hundred people will die.

Which of these do you pick?

For the first set of programs, 72 percent of the subjects picked Program 1 in the original study. For the second set of programs, 78 percent of the subjects picked Program 4. I'm sure you've noticed; however, that Programs 1 and 3 are the same, and Programs 2 and 4 are the same. The only difference is the way the information is framed. In the first instance of framing the program, 72 percent of people chose it; in the second instance of framing the same program, only 22 percent chose it. When framing blurs judgment, the nuts and bolts of the content matters far less, whether or not we realize it.

The concluding irony of the Ed and Ned story is that the best decision was to let the thieves go and call the police—which, incidentally, is exactly what I was doing. As I mentioned earlier, one thing we can say in defense of Ed and Ned is that they were acting under time pressure. But as we'll discuss next, screwing up with framing doesn't require an iota of urgency.

MISFRAMED IN TRANSLATION

A few years ago, I was on a business trip in China and had a day to take in some sights. Number one on my list: the Great Wall. With me was an American expat named Mark who had lived in Beijing for several years and was intimately familiar with local customs, and also, thankfully, spoke fluent Mandarin. He had been instructing me throughout the trip about the best ways to interact with the Chinese, things not to do, and things that I should not assume—even though in the States they'd be perfectly valid assumptions.

One of his firmer admonitions was that when approaching street vendors, be extremely careful not to touch anything you don't intend to buy. While doing so is normal in the States, the Chinese street vendors perceive your willingness to "investigate" their products as a sign that you will buy something. This particular rule struck me as ridiculous.

How would I know I wanted to buy something if I didn't at least pick it up to look it over? His response: "Just don't. Trust me."

Along the path leading to the Great Wall are several vendors selling what seems an infinite assortment of memorabilia. One item especially caught my eye: a vintage replica of a Chinese Red Army winter hat, complete with massive fur-lined ear flaps. I walked over to the vendor booth displaying the item. *In this case*, I thought to myself, *surely one would be expected to at least try on the hat to see if it fits before buying it.* Ignoring my colleague's advice, I picked up the hat and put it on. It didn't fit very well, so I placed it back on the display table.

Before I realized what was happening, two vendors from the table circled around behind me. One of them grabbed the hat and handed it back to me while emphatically chattering in Mandarin. While I couldn't understand anything he said, the implication was clear: Buy the hat. I smiled and shook my head, trying to highlight my foreign naïveté, and put the hat back on the table. Then the other vendor grabbed it and this time forcibly put it on my head. I pulled it off and put it on the table, only to have it once again put back on my head. Eventually, Mark saw what was happening and intervened, but, even then, we were pursued by the vendors until they gave up chase near the entrance of the Great Wall. This framing foible is of a different flavor than Ed and Ned's. I was given explicit, credible information that behaving in a certain way would result in a bad outcome. I also had enough time to fully consider my actions. Yet I still framed the situation in line with my narrowly defined criteria and acted accordingly.

Research by psychologists Keith E. Stanovich and Richard F. West suggests that my error in this case was a failure to engage "heuristic override."[11] In psychology, heuristics are simple, efficient rules—either hardwired in our brains or learned—that kick in especially when we're facing problems with incomplete information. Heuristics are terrific tools that humans rely on all of the time, but it's always quite possible that they are leading us down a blind alley—which is precisely when we should allow outside knowledge to supplement, and perhaps trump, our inner leanings.

When skewed heuristics win the day, psychologists like Stanovich and West tell us we've succumbed to *heuristic bias*. One way to understand heuristic bias is to imagine a rule book filled with (1) rules that have always been in the book, and (2) rules that are always being added to it. You keep the rule book with you at all times and refer to it often. There are, however, a couple of rather big problems with the book: Many of its rules are written as if they are absolute and irrefutable (though they are neither), and many are written as if they apply to all situations (they don't). Unfortunately, it's hard to tell when a rule shouldn't apply, so you often find yourself deliberating about what to do. The more you question the rules, the more ambiguity you face, and the more your brain feels threatened. The tendency, then, is to go with the rule and settle the matter. That makes your brain happy, though the result might be anything but pleasant.

Reflecting back on the story, I started by framing the vendor situation narrowly, to the exclusion of information presented to me. I then acted on a rule from my book: Always disregard advice if it sounds inconsistent with my experience. I felt "right" making those decisions, both of which turned out badly with more than a little awkwardness to boot.

We'll discuss more about knowing when to press the override button at the end of this chapter. For now, let's leave framing and heuristics and take a quick trip to antiquity (sort of).

THIS IS SPARTA! HIKE!

I'm a big fan of the movie *300*, a vivid retelling of the Spartans' famous stand against the Persians at Thermopylae—where three hundred of

Sparta's best warriors held fast against enemy numbers too vast to estimate. More precisely, it's a re-creation of comic book icon Frank Miller's graphic novel about the Spartans' fight "for freedom" against the notoriously cruel Persian army.[12]

Why I like the movie has nothing to do with historical accuracy, of which it has quite little. Aside from enjoying its well-choreographed fight scenes, I like the movie because it's an excellent example of how information is chosen to confirm existing positions.

Shortly after its release in theaters, certain right-of-center print, radio, and TV personalities picked up the film's themes of "fighting for freedom" and "standing against oppression" and integrated them into their shticks. The movie became part of their histrionic populist anthems—and its buffed and tanned protagonists, the Spartans, became populist heroes.

What the same media personalities did not roll into their monologues is that the Spartans were also brutal slave owners. The reason that Spartan men could devote themselves to becoming full-time professional warriors is that they had plenty of slaves for tending mundane, nonwarrior labor. The Spartans were not concerned with the principle of "freedom"—they were concerned with their own freedom.

The Spartans were also hardly the sort of heroes anyone preaching "family values" should venerate. Spartan babies were routinely thrown off cliffs for not displaying the qualities valuable to a warrior nation. Spartan boys were left to fend for themselves at an early age, often starving to death or being killed in a style of training barbaric to modern sensibilities. This viciously utilitarian perspective on life is the antithesis of what most Americans claim to embrace.

The question is: Why didn't any of that matter? How did the themes of a movie built with fictional Popsicle sticks become so important in the family-centric, populist milieu? The answer is that the theme of freedom, and everything that supports it, fit into the existing beliefs of those preaching and subscribing to the populist message. The historical realities that call into question almost everything about that theme did not fit—and, hence, were excluded. Psychologists refer to this tendency to seek confirming evidence and ignore disconfirming evidence as *confirmation bias*, and it's as human as sex, sleep, and barbeque.

Another quick example involving burly, violent men will further illustrate how this works. Let's say that you are a die-hard fan of the New York Giants. One day you are online and decide to surf around a few professional football message boards to see what the latest buzz is, and you come across a thread titled, "Why the Giants Suck: A Definitive Argument." You click on it and start reading the entries. Some of them are blatant insults, and while those make you mad, you don't pay them much attention. The posts that really grab your attention are those that call into question certain facts about the team that you, as a longtime fan, have taken to be true without qualification. Further, a few of the posts aren't just inane one-liners but well-thought-out arguments with supporting evidence.

What are you feeling right now? It's not exactly the same as anger, because anger alone is a shotgun force. Your emotions are more directed and mixed. While you are mad that people are slamming your team, you are also concerned that what some of them are saying might actually be true. And if some of it is true, that means some of what you've held near and dear is false. "That can't happen," you say to yourself, "but I also can't just ignore what I've read." So you determine to do some research and write a post of your own, and you know exactly where to start: Giants fan sites. Where else would you find information to build your case?

Stop tape. You see where this is going. Your motivation in this case isn't to rationally evaluate the information you've found and, if necessary, change your position. Your real motivation is to find information that will confirm your existing position, because there's no way in hell that it's changing, period.

We do this all of the time with all sorts of beliefs and positions. What is it we're really after when we seek evidence to confirm, and dismiss or ignore evidence that disconfirms? In a word: closure. More on that after a visit to Turin and a chat about your dog.

THE SHROUD OF TURIN AND WASHING THE DOG

The Shroud of Turin, that fantastically resilient white whale of religious archeology, is never far from a new round of debate about its authenticity. What makes its resurfacing remarkable is that every credible scientific evaluation to date has exposed it as a medieval hoax. Repeatedly, radiocarbon dating has shown that the shroud is a thirteenth- or fourteenth-century creation, not a first-century phenomenon. Yet the debate persists—and with vigor.

In a 2009 volley from the the-shroud-is-real camp, a Vatican researcher claimed to have found virtually invisible words written in Greek, Latin, and Aramaic across the shroud. The words, she claimed, include the name "Jesus the Nazarene" in Greek. If the shroud were a medieval forgery, the contention continues, Jesus' name would not have been used without reference to his divinity. Even a smarmy forger wouldn't make such a blatant error, the Vatican's researchers argue; hence, the shroud must be from the first century. The lead researcher was a historian employed by the Vatican and known for her studies of the secret order of the Knights Templar. During that research campaign, she claimed that the Knights at one time had the shroud in their possession—which strikes most historians who study the period as odd since the Knights were disbanded in the early fourteenth century and the first record of the shroud's existence occurred around 1360 in France.[13] Her more recent contention about the shroud was also met with hefty skepticism. For years, those who have studied the shroud have known that it is peppered with words—nothing new about that—but she is the first to claim that she's found definitive proof of the artifact's authenticity within those words. There are, however, a few insurmountable problems with her conclusion. For example, she claims that part of the text is written in Latin—a language never used during Jewish burials in the first century. Deal breaker? Not in this debate.[14]

How Uncertainty Makes You Lose Your Mind

I'm going to give you three options. You'll be OK with only one of them, and one of the other two is going to nudge you a little closer to insanity. Ready? Here they are: (1) I am not going to give you an electric shock, (2) I am definitely going to give you an electric shock, or (3) I might give you an electric shock. While it goes without saying which one you're OK with, which of the other two is starting to make the skin on your forearms itch a little? According to a study probing the effects of uncertainty on the brain, it's probably not option 2 (though it's also not going to make me any friends)—rather, it's what's behind door number 3 that's stirring up the bats in your belfry. When we're facing outcomes imbued with uncertainty, it's the fact that something bad *might* happen that gets us. Researchers recruited forty-five volunteers to play a computer game in which they turned over digital rocks that might have snakes hiding underneath. Throughout the game they had to guess whether each rock concealed a snake, and when a snake appeared they received a mild but painful electric shock on the hand. Over the course of the game they got better about predicting under which rocks they'd find snakes, but the game was designed to keep changing the odds of success to maintain ongoing uncertainty. In other words, they could get only marginally good at guessing before getting knocked off their game again and again, and again. In the background the researchers were running a sophisticated computational learning model to estimate the volunteers' amount of uncertainty that any given rock was concealing a snake. At the same time, their stress was being monitored via instruments tracking pupil dilation and perspiration. As it turned out, the volunteers' level of uncertainty correlated in lockstep with their level of stress. If someone felt "certain" he or she would find a snake (100 percent probability that a snake lives here), his or her stress was significantly lower than if he or she felt like *maybe* a snake lives here. In both cases the volunteer would get a shock, but the added uncertainty caused the most profound stress reactions.[15]

The question is, why do people dedicate themselves to proving something true that has already been proven false? First, we have to appreciate the depth of emotion tied to this belief. Despite appearances, this debate is not solely about evidence. For many, a commitment to the shroud's authenticity is integral to their overall spiritual beliefs.

If the religious belief position cannot be separated from a closer-to-objective scientific evaluation, then the outcome will always be the same. The stakes are too high for disconfirming facts to prevail; the evidence will forever be in question, no matter how substantial.

Taking the discussion down a hundred or so octaves, let's pretend that you and your significant other are arguing about whose turn it is to wash the dog—easily one of the least pleasant chores either of you have on your to-do lists. You both believe you were the last one to do it. But the fact is that you both are also very busy people, and your days and nights tend to blend into each other. Either of you could easily be wrong about who washed poochy last.

Yet you stand your ground, arguing your position as if you were addressing the United Nations. But then your partner hits you with strong evidence challenging your position. You could not have washed the dog on the day you said you did because you had to leave early for work that day to attend a meeting. All along you've maintained that you got up early to wash the dog—both can't be true. The argument sounds plausible, but you're still sure that you are right. Perhaps you guessed the day wrong; it was, after all, a couple of weeks ago. Anyone could make that mistake.

But the armor has been pierced, and now you are starting to wonder if you really have your story straight. If you can't recall which day you had an early meeting, then it's only reasonable to conclude that you could forget which day you, or someone, washed the dog. Is it possible that you've been selecting the scraps from memory to make your argument? Is it probable that your perception has morphed to fill the shape of your expectation—that you shouldn't have to wash your much-loved but extremely dirty pet? With evidence stacking against you and your certainty losing ground by the second, will you give in?

The Shroud of Turin flavor of confirmation bias is linked to a deeply embedded religious belief. It's similar to the debate over evolution in that giving any ground in the argument is tantamount to disavowing one's heartfelt spiritual beliefs, the standing of which trumps all else. The washing-your-dog flavor of confirmation bias is not belief-based, but it is linked to a sense of personal credibility—your "track record" of being right or wrong. For most people, a credibility position is not as unyielding as a belief-based position, but it's a stubborn beast nonetheless.

We can reasonably expect that, eventually, someone will give on washing the dog (after all, the dog won't wash himself), though that doesn't necessarily mean someone will admit that she or he is wrong. Even with something so trivial, confirmation bias is a tough force to over-come; both parties want the cerebral surge that comes with being proven right. Arguments like washing the dog offer what you might call "low-hanging fruit." Getting closure on relatively small matters is like picking off the easiest fruit to reach on the proverbial tree. Our brains like this because each instance of closure is a little reward—a tasty jolt of cer-tainty. If it were possible to "win" the Shroud of Turin debate (it's not), the resulting surge would be epic. Gaining cognitive closure at that level is, to use another reference from antiquity, the Holy Grail of certainty for our brains. And in the case of the shroud, that certainty will, like the Grail, always remain out of reach.

But what might happen if a deeply held belief–based position were proven wrong in such a way that no reasonable person could challenge the outcome? Would the stanchions of confirmation bias crack and col-lapse under the weight of overwhelming evidence? Let's see.

THE KIAI MASTER'S CHALLENGE

A number of martial arts sects across the globe claim to teach an uncanny ability known as the "touchless attack." It is exactly what it sounds like: the ability to knock opponents on their derrieres without physically touching their bodies. One practitioner of this invisible assault goes by the moniker "The Human Stun Gun." Others call themselves "Kiai masters" (*Kiai* being a spelling variation of the term *Chi* or *Ki*, referring to one's inherent inner "force" that is allegedly harnessed via certain martial arts techniques).

Make no mistake—those who believe in the touchless attack are resolute that it is real. Videos of Kiai masters effortlessly flinging their students around their dojos (training facilities for martial artists) are legion on the Internet. With nothing but swift hand movements through the air, these masters flip, toss, trip, and knock out each student as he or she attempts in vain to attack the carnage-wielding magician before them.

Outsiders—those not students of the touchless attack—question its validity (critics of the practice call it "*bull*shido"). Many have challenged the Kiai masters to demonstrate their ability with opponents who are not their students, and usually these challenges are ignored or rebuffed. They have nothing to prove to outsiders, the masters frequently respond. The "proof" is right there in the videos of their sessions for all to see; take it or leave it.

One Kiai master, however, determined that not only could he prove his abilities with nonstudents, but he'd even be willing to put money on it. The wager: $5,000 to anyone who could face the Kiai master and withstand his lethal assault in front of Japanese television cameras. To make the wager even more interesting, he issued the challenge directly to mixed martial arts (MMA) practitioners—those skilled in multiple arts like karate, jujitsu, and kickboxing. Not surprisingly, someone took him up on it.

Kiai master Ryukerin, who claimed a 200–0 record coming into the match, faced his opponent in his dojo with a live crowd and television cameras rolling, per the terms of his challenge. His opponent, an experienced mixed martial artist, publicly signed the contract before the bout, agreeing that if he could not withstand Ryukerin's assault, he would not win the $5,000. With all formalities completed, the pair began the match.[16]

It didn't last long. The MMA fighter charged at Master Ryukerin—who moved his hands through the air in the same manner that flipped his students on their heads—and mercilessly pummeled the Kiai master as the crowd looked on in horror. Ryukerin got up and again waved his hands about in the general direction of his opponent, and was again knocked to the ground and repeatedly kicked until finally submitting.

Anyone watching the video of this fiasco is forced to wonder why Ryukerin would willingly subject his belief (and body) to such open and painful scrutiny. Whatever the answer, what seems clear is that he genuinely believed that he possessed an ability to move physical objects without touching them; he put his money, his reputation, and his pride on the line to prove it. In doing so, he unwittingly provided us with an incredible example of belief-based confirmation bias facing the ultimate challenge—and losing big. I'm not talking about *his* confirmation bias, mind you; I'm talking about the confirmation bias of those who believe what he believed—his students and other supporters who fervently argued that the ability claimed by the Kiai master was 100 percent legitimate. Surely they could no longer stand on this argument, right?

Wrong. In the aftermath of the fight, many supporters of the touchless attack didn't move an inch. The MMA fighter had somehow managed to "channel" the energy of the master's attack, they argued. Ryukerin was under the weather and unable to wield his power at full strength, said others. The loss was an anomaly. It proved nothing. The touchless attack and those who teach it did not go extinct as one might expect after thousands upon thousands of people watched the brutal uncloaking of Master Ryukerin.

That, in a nutshell, is the power of confirmation bias. You can punch it, kick it, break its arms and legs, and humiliate it for all to see—yet still it stands.

SCHEMA FOR YOUR THOUGHTS

Taking a position in any argument—large or small—is slippery business for our brains. We can have every intention of honestly pursuing an answer, yet still fool ourselves into thinking our method is objective when it is, in fact, anything but. Cognitive science has helped decipher this enigma with research on the theoretical mental structures our brains use to organize information, called *schemata*.[17]

A schema (singular form of *schemata*) is like a mental map of concepts that hangs together by association. For example, your schema for "school" contains associations between "teacher" and "books" and "subjects." Each of those have additional associations; "subjects" is linked to "math" and "literature," for example. Cognitive science suggests that as schemata develop, the parameters for what information can be included tighten. The reason for this is very practical: We make judgments based on the linkages in our schemata. If the information didn't hang together in a structured way, and if certain pieces of information were not excluded from the map, we'd find making even basic judgments extremely difficult.

Imagine that you've been in the workforce for about ten years and are interviewing for a job. The interviewer tells you about the job's duties, the work schedule, the location, the wage, and other pertinent details. All of this is important, but what's equally as important is what you brought into the room with you. Your schema for, let's call it "career," includes a host of linkages that have developed with time that you draw upon to make judgments. Is the company you are interviewing with compatible with your career? Does the schedule fit, does the wage fit, does the size of the company fit, does the commute time fit? You may reasonably change your mind about any of these things, of course, but the point is that you did not enter the room as an empty bucket ready to be filled. You entered with a preestablished schema for "career" that serves as the platform for your judgments.

And therein lies the rub. Preestablished schemata guide our attention to evaluate new information, but they can also guide our attention to selectively ignore information inconsistent with the schemata.

To understand why, we have to go back to what makes the brain happy. When a well-established schema is called into question by new information, the brain reacts as if threatened. The amygdalae fires up (threat response), and the ventral striatum revs down (reward response). This is not a comfortable place for the brain. The supercharged clay in your head doesn't like being on guard—it likes being stable. Ambiguity, which might result from considering the new information, is a threat. We can either allow that threat to stand by considering the inconsistent information, or block it by dismissing or ignoring it. Or we might subcategorize the information and store it away as an "outlier" case; something that can't be entirely ignored, but does not challenge or change the existing schema.

Cognitive-science researchers are especially interested in how our brains maintain preestablished schemata. Successfully plumbing the depths of religious belief, for example, appears to hinge on understanding the ways our brains seek stability. Indeed, belief in general appears to have much to do with the brain's penchant for homeostasis—defined by renowned physiologist Walter Bradford Cannon as "the property of a system that regulates its internal environment and tends to maintain a stable, constant condition."[18]

We humans are prone to divide belief positions by value. Believing in God is more important than believing 2 + 2 = 4. But neuroscience research has shown that in the brain, all belief reactions look the same, whether the stimulus is value-laden (like religion) or neutral (like math).[19] Whether the value we've assigned to a belief is—from our subjective vantage point—high or low, the brain wants the same things: stability and consistency. We seldom realize it, but very nearly everything we do is colored by this drive.

ENGAGING COGNITIVE OVERRIDE, CAPTAIN

A new product is about to hit the market, and I think you'll want to take notice. It's called the "Super Novum." Shaped like a slightly over-large motorcycle helmet, the user places it on her head and pushes just one button to get things started. She doesn't know it yet, but she has just given her brain an amazing advantage over all of the other brains walking around out there. Some of the features she'll experience include greatly reduced selective attention—no more missing the details! Broader framing—no more mental myopia! And information that challenges her beliefs can drive on in for an objective evaluation—no more confirmation bias! Plus, the Super Novum comes in a variety of colors and patterns to match its user's unique personality.

Even if such a device existed, I wonder if we'd really want it. Would it be worth short-circuiting parts of our brains to avoid the sorts of certainty foibles discussed in this chapter? Probably not. A better question might be, if the brain craves certainty, then why not simply give it what it wants? Why not abide the urge to feel "right" if that's what makes the brain happy?

Before I try to answer those questions, I want to tell you a brief story about Jennifer, who likes jumping out of airplanes. At some point early in adulthood, she decided that her urge to leap from a perfectly stable plane had been put off long enough. So she found a reputable skydiving outfit to kick off what was sure to become a lifelong passion for death-defying sports. From my perspective, this was just short of insanity. "So you're going to step out of a plane at 12,000 feet?" I recall asking—as I glanced over the liability disclaimer forms (with statements like, "You acknowledge that engaging in this activity can result in your sudden death."). For her, every moment leading up to the jump was sheer ecstasy. Not that she wasn't nervous (only a zombie wouldn't have some nervous reaction before jumping from an airplane thousands of feet above sea level), but the exhilaration of doing what she'd wanted to do for so long—to take on one of her ultimate challenges—outpaced her anxiety by a furlong. She went on to have a successful jump, and I managed to watch the whole thing without closing my eyes.

Why Believe in the Unbelievable?

According to cognitive psychologist and author Bruce M. Hood, we may all be born with a "supersense" that leads us to believe in some very odd things. In his book *SuperSense: Why We Believe in the Unbelievable*, Hood argues that humans are born with brains structured to make sense of the world, and that often leads to beliefs that go beyond any natural explanation. To be true, they would have to be supernatural. "We are inclined from the start to think that there are unseen patterns, forces, and essences inhabiting the world. This way of thinking is unavoidable, and it may be part of human nature to see ourselves connected to each other at this deeper level."[20] Hood also suggests that these beliefs may operate to bind individuals together on the basis of shared sacred values that transcend the mundane by becoming "profound."

We have to appreciate that our brains weren't born yesterday. We have mechanisms to warn of threats and guard against instability because they have worked for a very long time. We wouldn't be here without them. In the same way that any sane person feels apprehension about jumping out of an airplane, our brain puts the organism it controls on alert when danger looms—be it tangible or intangible. But we have to know when to override the alarm and take the less comfortable path anyway.

Research conducted by a joint American and Italian team of psychologists found that people with less need for "cognitive closure" were typically more creative problem solvers than their counterparts.[21] In other words, those who are able to work past their brain's appetite for certainty—its need to shut the closure door to preserve stability—are more likely to engage challenges from a broader variety of vantage points and take risks to overcome them. Jumping out of the airplane even when our brain is shouting "Stop!" is sometimes exactly what we need to do. That's the energy that fuels scientific discovery, technological advances, and a range of other human pursuits.

Which is not to say, of course, that we shouldn't also listen to our brains. It's not always advantageous to act against our neural inclinations. Sometimes a narrow frame is right for the situation, and sometimes disallowing new information is necessary. We have to dance with our instincts to figure out when to leap or when to stay on the ground. That's the challenge of being human—of having a big brain capable of greatness with hardwiring evolved for survival.

Chapter 2

SEDUCTIVE PATTERNS
AND SMOKING MONKEYS

"There are as many pillows of illusion as flakes in a snowstorm."

—RALPH WALDO EMERSON,
CONDUCT OF LIFE

WHO'S TELLING ME WHAT?

Let's try a little thought experiment. First, imagine that you are walking through an airport. As you stroll, you encounter a series of variables that occur randomly, but could easily be interpreted as uncannily coincidental. For instance, the same number (let's say 429) appears in four different places in the span of about forty-five minutes: the price of a magazine; the time on your watch when you happen to glance at it; a number imprinted on the back of someone's T-shirt; and the cost of a frozen yogurt. The stroll, and these weird, rapidly occurring coincidences, is leading up to boarding an airplane. As it happens, the number of your flight fits eerily into the trail of coincidences: It's Flight 1429. You are faced with whether to read meaning into these coincidences or ignore them. Have you been given a "sign" about the flight? If so, by whom? And what could it mean? Should you get on the flight or change your ticket for the next one with a number that isn't part of the strange pattern?

An example like this is enough to challenge most people's skepticism about whether random occurrences mean anything when they appear to fit a pattern. But if the pattern means something, what's the compass for figuring out the meaning? Since you are about to board an airplane, the first reaction is that you have been given a warning not to get on the plane. Fair enough. But then, how many other people about to get on the same airplane received a comparable warning? And if they didn't, why are you the recipient of such privileged information? And then there is an entirely different possibility that you have been given a sign that something good is about to happen to you on the plane. Maybe you'll be seated next to your future spouse or someone who is going to eventually offer you a high-paying job. But, again, how can you really know for certain? Is it worth taking the risk? If you get on the plane and it crashes, as you descend, you will know that you could have acted on the "sign" but chose not to. Just thinking about that possibility causes beads of sweat to well up along your forehead.

OK. Now stop and take a breath.

If you found that paragraph anxiety-inducing to read, then the first part of my point has been delivered. It is simply this: The meanings we give to patterns of coincidence originate and live solely in our minds and are then projected into the world. The problem is that once our pattern-detecting brains are provided with the rudiments of a convincing story—one with possibilities that may endanger or benefit us—it is difficult to pull out of the process. In this case, it will be hard to rewind the mental tape and ask yourself why you started looking for 429 to begin with. Perhaps you saw the number twice, and even that minor pattern put your brain on alert, with the result of tightening your focus to extend the pattern. Once that happened, finding more 429s in the airport would not be difficult. They would begin sticking out like red neon signs against a black backdrop. The reality is that your safety was never any more in jeopardy than if you had never experienced the coincidences at all—and, rationally, you knew this all along. But as we will continue to discuss throughout this book, the gap between a happy brain's library of knowledge and our actions is larger than we know.

Here's another example, this one entirely true and pulled from my personal repertoire of chance experience. I was about eight years old, living in Rochester, New York, playing by myself in the snow near my driveway. Three odd things happened while I was outside. First, I smelled something foul, like a dead animal. At that age, I had not been around too many animals (aside from our dog), but for some reason I connected the smell in the air with an animal of some sort. It drifted off in just a couple of minutes, and I continued playing. Next, I heard what I thought was loud breathing—too loud to be a person. I looked around the house but didn't find anyone, or anything, and the noise soon stopped. Finally, while I was building a snow fort out of a snowbank along a walkway to the house, I looked down and saw a splotch of snow on the walkway with a distinct shape. The more I looked at it, the more defined the splotch became—and I concluded that the shape was that of a bear: head, body, legs, and claws. Not long after, my mother called me into the house for dinner. The local news was on TV, and my attention was immediately riveted by one of the top stories: A black bear had escaped from the city zoo and was still at large.

So what do you think? Random coincidence or a sign to avoid danger? I'll leave that for you to digest as we move on.

I THINK, THEREFORE I CONNECT

Since our brains are adept at finding and drawing conclusions from patterns, it's not surprising that coincidences captivate our attention. The pioneering psychologist Carl Gustav Jung went so far as to argue that what we take to be coincidences are actually interlinked associations that form a sort of unseen web (he used the term *synchronicity* to describe this ethereal force).[1] A cottage industry within both the New Age and self-help spheres capitalized on Jung's claims; the international bestseller about the power of coincidence, *The Celestine Prophecy*, and its follow-on books are prime examples.

This compulsion to connect experiences, symbols, images, and ideas stems directly from the brain's vital function as an organ evolved to make sense of our environment. As we have discussed, without this ability, our species would have vanished long ago. The problem is that our penchant for connection—like many features of our brains—can get out of hand. When that happens, our brains quite literally make something out of nothing, and we can't seem to stop ourselves from doing it over and over again.

Consider, for example, the true case of a psychologist who has become enamored with the mega-bestselling book *The Secret*—another refashioning of positive thinking and the "think yourself rich" mantra— and recommends that her patient read the book as well. As proof of how much the book has changed the psychologist's life, she shows the patient a photo of a new BMW convertible. She explains that the book convinced her that she must focus on that which she wants the most (the car) and constantly remind herself that she *will* own the BMW. She adds that ever since she started doing so, she sees the car everywhere. In fact, in the last few days she has seen it no less than five times on the road—and this, she states, is surely a sign that her positive-thinking strategy is working. Even as a trained psychologist she fails to see that she has duped herself. With her attention primed by an image of the car, her brain is busily picking it out of the scenery everywhere she goes. Instead of pushing through the muddled thinking, she concludes that her positive focus is producing "signs" that her goal will soon be reached.[2] If she eventually goes on to buy the car, she'll credit this pattern-driven ambition, as if a mysterious force was guiding her, when every component of the process was in her control from the start (and some would argue that's the real value of books like *The Secret*, whether or not many of its fans know it).

Even if you do not consider yourself especially pattern-prone, you probably still fall into the connection trap without even thinking about it, albeit in not so dramatic fashion as our earlier airplane example. Marketers routinely use something psychologists call the *clustering illusion* in retail product placement that leverages our compulsion to identify patterns and assign meaning. If, for instance, three Blu-ray disc players are placed next to each other with the highest priced item followed by the next highest priced item, followed by the lowest priced item—then the store can expect to sell more of the item in the middle, and will mark it up accordingly. The reason is that we assign a specific meaning to the placement (best, next best, worst), when it's possible that no such meaning exists. In fact, the Blu-ray player in the middle may be of no higher quality than the cheapest player. Retail marketers know that people are convinced that their meaning associations are correct, and they leverage them to reap a higher markup on cheap items.

Our brains make these associations because that is what brains do (in part), and it's always helpful to remember that the adaptive capabilities of our brains—like pattern recognition—did not evolve to make sense of complex commercial environments like those we live in.[3] As we've seen, and will continue to discuss, we've developed modern cultures that are in many ways at odds with our brains' capabilities. Making things even more interesting, we are also predisposed to wanting an evident cause for every effect—and if one isn't evident, our brains will happily create one.

ALLEGED CAUSES, PRESUMED EFFECTS

Imagine that a study on the effects of drinking coffee comes out in the news indicating that drinking at least three cups a day significantly improves attention and memory. Maria reads this news and finds it convincing, so she increases her morning coffee ritual to three cups. For the next month she thinks she is more attentive and remembering things better because she's drinking more coffee. Then she reads a newer study that says drinking more than two cups of coffee a day is linked to significantly decreased attention and heightened anxiety. The second study has been promoted with as much gusto as the first, and the credentials of the researchers are just as impressive. She thinks, *You know, I have been feeling more anxious lately, and maybe I'm not as focused as I thought*, and she decreases her coffee intake down to two cups.

For the next couple of weeks, she feels more attentive and not nearly as anxious—until she reads an article a few weeks later discrediting the second study and upholding the findings of the first. At each step along the way, the effects Maria was experiencing had far less to do with coffee and far more to do with her belief that causal links existed between a behavior and an outcome. Daily we are barraged with examples like that (health claims in particular, because popular media blasts them out into the world as soon as they're published or reported by researchers), with the result of feeling like we're on a perpetual seesaw of pros and cons.

My analogy of choice for highlighting this tendency is the "smoking monkey," referring to a kitschy novelty item popular in the 1960s. A small plastic or ceramic monkey is packaged with a tiny cigarette that is placed into a hole in its mouth, and when the cigarette is lit the monkey appears to smoke, even blowing smoke rings. The monkey is hollow, and there is a second hole in its bottom, allowing air to circulate through its body and keep the tiny cigarette smoldering. At least, that's the cause-and-effect takeaway someone might have when the hole in the bottom is shown to them and the air circulation link is suggested. In much the same way, when a person smokes, air pumped from the lungs, circulates through the cigarette, and keeps it burning.

That would be a neat and tidy analogy that makes sense of the original puzzle—except, the smoking monkey does not work at all the same as a smoking human, and the explanation is wrong. In truth, the hole has nothing to do with the cigarette—which is instead made of paper designed to smolder without burning. We are usually like the person who sees the other hole, gets a little information, and concludes a cause-and-effect link (what psychologists call *causation*). We're faced with smoking monkeys every day, and our brains are happy to fill in the blanks with causal relationships that don't really exist. What our brains are searching for in each smoking-monkey scenario is a "story" that makes sense. Next we will discuss why.

IT MUST MEAN SOMETHING, RIGHT?

Storytelling is powerful medicine for the mind. One of the reasons stories appeal to us (in books, on TV, online, or otherwise) is that they link together shards of meaning that eventually yield even greater meaning. In other words, stories make sense of the world. Making sense of the world makes our brains happy. But some of the stories we hear lack an adequate wrap-up. Here's a true story that illustrates the point.

A few years ago I was working on a public health campaign in Birmingham, Alabama, and heard some news that put a tragically capital *R* in *Random*. A woman driving downtown stopped at an intersection and waited for the light to change. What she didn't know is that she had stopped her car directly over a water main manhole cover. What she also didn't know, and could not have known, is that the city was experiencing a massive pressure surge in the water main, which was building in intensity as she approached the intersection.

In the handful of minutes that she waited for a green light, the pressure surge reached the part of the water main where her car had stopped, and—having hit the weakest part of the pipeline—erupted as a geyser of scorching hot steam through the manhole. She was steamed to death in her car like a lobster in a pot of boiling water.

It's difficult to imagine the odds of such an exceptionally random event, but I did some rough figuring and came up with about 1 in 500,000 (taking into account the average number of drivers in downtown Birmingham at that time of day, the number of manholes, and the chance of that sort of water-pipeline problem happening; I later learned that it's called a "water hammer"). I'm sure my figures are far from perfect, but whatever the actual number is, there's no question that the chances of dying that way are remote. And yet, on one idle afternoon when everything seemed just as normal as any other day, it happened. Upon hearing a story like this, we can actually "feel" how our brain wants to string together the chance events leading up to the outcome, in an effort to make sense of the tragedy. But even with a physical explanation as to why it happened (pressure surge), the story lacks closure at the level of *Why* (capital *W)* it occurred.

The reason this open-endedness is hard to accept is because it reinforces the sense that random tragedies can happen to anyone, including us—and that is mighty threatening to a threat-sensitive brain. Said another way, the lack of a *Why* underscores the power of randomness in our lives. We crave a reason. Hence the oft-quoted statement, "Everything happens for a reason." What is the reason? We don't know, but asserting that there *must* be one acts as a surrogate for closure. It also provides us with something absolutely necessary for a reason to exist: agency. An "agent" in psychological literature is a person or thing responsible for causing something to happen. We search for agents all of the time—personal and impersonal—and we select words that imply agency even when we know it doesn't exist. For example, a professor is attempting to give a presentation to his class using a computer and a projector. The projector isn't working, and after several attempts to fix it he says, "It seems this projector is determined to wreck my class." He knows, as does everyone in the room, that the projector is not an action-causing agent, but his words betray the brain's desire to assign agency no matter the physical facts. We blame our car for not starting, software for not saving documents, plants for not growing, and on and on. The philosopher Daniel Dennett calls this the *intentional stance*: we refer to objects both animate and inanimate as if they have minds, as a shortcut to figuring out what is really going on.[4]

Again, we can find a likely evolutionary underpinning for this tendency of a happy brain—namely that identifying what is causing an action could save our lives. Picture one of our ancestors gathering food in the thick of the forest. Suddenly he hears a rustling in a nearby tree. Is it the wind, a harmless bird, or a massive man-eating cat? Decoding the clues quickly and finding the actual cause could be the difference between returning to the family with dinner or becoming dinner. Leaving the forest, we can also see how this tendency would evolve for deciphering the intentions of others in more modern contexts. The human animal is the most formidable on the planet not only against other species but also against other humans. Not correctly identifying another's real intentions could very well be the last mistake a person makes.

Fallacy of Conjunction, What's Your Paranormal Function?

The fallacy of conjunction refers to the mistake of thinking that the "conjunct" of a true statement is truer than the statement itself. For example, if I tell you that Jim is an elected official and that he is an avid target shooter, you will instantly develop an image of the sort of person Jim is. If I then ask you to select which statements about Jim is truer, that (1) he is a politician, or (2) he is a politician who doesn't support gun permitting, you are faced with two statements, both of which encapsulate the one true thing we know about Jim (he's a politician). The second statement, however, includes an inference that because Jim is an avid target shooter, he must not support gun permitting. Is that statement, then, truer than the first? No, it isn't, because even if you are correct about Jim's position on gun permitting, the first statement is still no less true. More important, the second statement is built on speculation about Jim's politics, and there is really very little reason to consider it accurate. Psychology research indicates that believers in the paranormal are especially susceptible to the fallacy of conjunction. This makes sense if you replace the true statement in the example above with a statement related to, let's say, a psychic cold reader accurately naming someone's dead relative during a "communicating with the beyond" session. If the cold reader gets the name right (and therefore the statement is correct), then he has an open door to add conjuncts to the statement. If the recipient of the cold reading falls for this, she is really falling into the fallacy of conjunction. Once she does, the cold reader has swindled de facto permission to come up with as many conjuncts as he likes—building a castle of falsehoods from a single truth.[5]

STATS AND YOUR BRAIN—NOT A LOVE STORY

Statistics is not most students' favorite subject. Next to calculus and organic chemistry, it might be the most avoided class in any college undergraduate program. The truth, however, is that statistics lord over our lives every minute of every day. For the purposes of this discussion, suffice it to say that all of us are the pawns of probability. Given enough time, enough drivers, enough incremental problems in the pipeline— eventually someone will stop his or her car over a manhole cover that is about to blow. It may happen only once in a year, or ten years, or maybe more, but it will happen. How do we know this? Because it happened.

The words *random* and *luck* are really stand-ins for a more jargony term: *probabilistic outcomes*. When a tornado rips through a town and destroys every house in a neighborhood except for one—which is somehow left intact right down to the manicured hedges—it is acceptable shorthand to call the house still standing (and the people in it) "lucky." Or someone might attribute the saving of the house and destruction of the others to an otherworldly agent like God or Satan.[6] But a diligent statistician would simply call it a probable outcome given the factors at play, such as the structure and speed of the tornado, location of the house compared to the others, and so forth. The statistician would also refer to a body of statistical evidence about tornado damage gathered over time that shows how frequently one house in a given neighborhood is left standing—and, conversely, how frequently only one house among several is destroyed. None of this information is going to precisely answer *Why* one house was spared or destroyed, but it will provide context for understanding that this event is not beyond explanation.[7]

Nevertheless, our brains have a hard time accepting the explanation. The immediate need to assign agency for what happened is strong enough to overpower the realization that many things happen without a *Why*. Indeed, on a less epic scale, they happen all of the time. I am still shocked every time I hear a word or phrase—for instance, a radio host says something about a "brown horse"—and I happen to look out of my car window and simultaneously see a brown horse in a stable near the road. Probabilistically speaking, this is not so remarkable. But in the moment it happens, I find myself looking for significance in the connection. It's not a silly mistake or magical thinking—it's what our brains do.

A FINAL WORD ON WHY THIS MATTERS

There is another dimension to this discussion that I have intentionally held for last in hopes that it might serve as a useful takeaway. Psychologists use the term *illusion of control* to describe what happens when we place ourselves in the role of agent in a situation that truly lacks one.[8] We tend to assume the role when something tragic happens to us or someone we love, and we think "If only I had . . . then this wouldn't have happened." In most cases, the control we think we could have exerted to prevent the tragedy is illusory. But the need to explain the *Why* of the tragedy, and a craving for agency will make it very hard for anyone to convince us that we are not in some way responsible for what happened.

Another way this plays out is in gambling—from state lotteries all the way to Vegas. Lotteries rely on the illusion of control for their very existence. Many lottery players are convinced that the numbers they have picked (as opposed to those randomly generated by a machine) are "better" numbers because they were selected by the player. If the player skips a day without playing those numbers and they come in, he is probably going to find the nearest ledge to jump from. So he plays his hand-picked numbers over and over and over again. He is operating under the illusion that he has control over the probability of the numbers coming in, but in truth he has not changed the odds of winning at all. If the numbers happen to come in, they'd have done so whether or not he chose them.

Casinos take advantage of similar thinking. The next time you go to a casino, ask a few people playing slot machines how they plan to win it big.

What you will hear from some is that they have a "system" for winning the slots, and as long as they stick to the system they will eventually win big money. Again, they are operating under the illusion that they can positively change the outcome with a formula for success. Unfortunately for them, the only real formula for success is to stop playing.

Why the House Always Wins

A study published in the *Journal of Gambling Studies* showed something quite paradoxical, but understandable, about online gambling. Researchers analyzed twenty-seven million online poker hands and found that the more hands players win, the less money they collect. The reason, the researchers believe, is that multiple wins usually yield small stakes—but to get those wins, you have to play long enough, and the longer you play, the more likely you are to suffer occasional big losses. Turns out, those occasional losses are enough to negate and often exceed the wins. Online gamblers misjudge the variance and uncertainty of the payoffs they get from taking risks, mainly because multiple small wins artificially inflate the players' sense of success. The result: The house always wins. It's a statistical guarantee voided only if you're one of the fortunate few who takes a big pot and calls it quits right then and there. Play on, and the house has you again.[9]

Part 2

DRIFTING, DISCOUNTING, AND ESCAPING

Chapter 3

WHY A HAPPY BRAIN
DISCOUNTS THE FUTURE

"I never think of the future—it comes soon enough."
—UNKNOWN

Your supervisor calls you one morning and tells you that she is heading up a new initiative. She describes in fine detail what this initiative entails, why the company is allocating a budget for it, and each of the expected outcomes. Eventually, she comes around to the question you thought might be lurking from the second you answered the call:

Would you be willing to take a major role in the new effort?

The problem is that your docket of projects is already teetering on unmanageable, and that situation won't be changing anytime soon. You explain this to your supervisor, who tells you that she understands—but, she adds, what she'll need you to begin doing on the new initiative won't kick off for at least six months. It is your choice, she emphasizes, and if you decide not to participate, it won't be held against you. The work you do on a daily basis is valued, and that will not change.

However, you think to yourself, if I do take on the new role, I'll be that much more valuable in the organization. And since the role won't start for at least six months, it seems like passing on this opportunity might be a mistake. So you say yes, you'll commit to taking on the new role in addition to your existing work. You end the call feeling good about the impression you've just made, and go about your work as you normally do.

A little more than six months later, you receive an email from your supervisor with a long, bulleted list of assignments—all due within two weeks—that you are expected to complete as a key participant in the new initiative. Your existing workload is still immense, as you predicted it would be six months ago, and now you have an overwhelming responsibility on top of it. The panic alarms sound and you mentally flog yourself for taking on the new role half a year earlier when you knew you'd be swamped. What were you thinking?

Whenever we are presented with a commitment that is a long way off, our normal tendency is to downplay the commitment—particularly if an immediate reward is involved.[1] In the preceding story, the immediate reward was a favorable reaction from the supervisor, which the employee believed would make him all the more valuable an asset. But when the commitment materialized, the employee's fear that he would be overloaded was proven out. Now what? His performance across all of his projects is likely to suffer because he said yes months before.

FUTURES UNCERTAIN

When presented with distant commitments, we stumble on the difficulty our brains have placing us in the future with any degree of accuracy. Because our brains evolved to make determinations about our existing environments and predict immediate threats and rewards, it's a stretch to gain perspective even a few months into the future. As important, our brains are always quite happy to capitalize on an immediate reward. When combined, the challenge to gain future perspective and the desire for immediate rewards sets us up for a range of problems. Economists call this tendency *hyperbolic discounting*.[2]

People selling high-ticket items leverage these tendencies all of the time to sell cars, houses, time-share units, and the like. When you are haggling with a car salesperson, take note that the figure she or he will focus on is the monthly payment. If you attempt to move the conversation away from the monthly payment, note that the salesperson will try to pull it back to that point. The reason is that if the more immediate issue—what you will pay every month—becomes palatable, the long-term issue—what you will be paying (including interest) years from now—will be overshadowed. So if you really can't afford a particular car, the salesperson's goal is to "put you into it" via other means; namely, by finagling the monthly cost over time to make you feel good about the purchase. The salesperson has a toolbox of ways to make this happen, and he'll keep trying tools until one fits your situation.

Note, also, that the financing of the car is handled in another room, by another salesperson. That person's job—adjunct to the first salesperson—is to maximize the car dealership's stake in your purchase while keeping you hooked. So, again, if you really can't afford the car, perhaps instead of a five-year loan, you'll be placed on a six-year loan. Whether you agree to five or six years doesn't seem like a very big difference from where you are standing right now. That you will be paying a full twelve more months of interest—potentially thousands of dollars—doesn't outweigh the immediate reward of driving off the lot with a brand-new car. The second salesperson's job is also to plug as many "products" (such as warranties) as possible into your sale and stretch them out accordingly to keep you focused on what you want in the short term. And above all else, the salesperson must keep you there to make the sale *that* day—for the simple reason that taking additional time to consider the long-term commitment will darken the rosy hue of the short-term reward. Selling is a game of momentum that leverages your brain's penchant for immediacy to close the deal, whether or not it was truly in your best interest.

Perhaps the most salient example of this tendency in action is something that frequently happens among friends and family members, and, unfortunately, often injures those relationships. Someone asks a friend or family member for help—let's say moving from one city to another. The move won't take place for several months, but it will require at least two full days of nonstop work to complete. In the moment the request is made, the person being asked for help likely wants to please the other, or perhaps feels obligated to say yes. In either case, the brain of the person being asked is seizing upon an immediate reward: gratification from satisfying the other person's need for help, or gratification from escaping the friction of evading a sense of obligation (this is an important point, because we usually think of a reward as "getting" something, but it can also be derived from avoiding something unpleasant).

The issue, of course, is not that the person being asked to help shouldn't want to help, but that it's very easy to overcommit oneself in the moment, and the consequences of overcommitment may be far worse than saying no up front. When the commitment eventually becomes a present reality, we're often left wondering how we could have committed ourselves when we have so many other competing commitments vying for our time and attention.

Evolution isn't directed by the interpersonal concerns of humans. Unlike our primate cousins, whose world is much more straightforward than ours, we have the added challenge of tailoring our responses to our expansive and complicated social environment. We place value on honoring our commitments, and failing to do so is viewed as a debit against one's character. What made our brains happy at the front end of a commitment tunnel may very well be what hurts us, and others, at the other end. All the more reason that we exercise a sort of cerebral restraint before satisfying the initial urge to collect our reward, in whatever form it takes.

DEATH BY URGENCY

Another way this tendency plays out has nothing to do with pleasing others (or feeling gratified from escaping the friction of "no"), and everything to do with clearing our mental and tangible desktop. Generations of corporate denizens (to pick on just one portion of society) have fallen victim to the syndrome of urgency that sacrifices sound, long-term decision making in favor of a "get it done" mentality. Faced with multiple deadlines and a lack of resources to address them all, it's easy to make quick decisions on matters in which the outcome won't materialize for a period of time long enough to mollify our anxiety.

Consider the example of a junior staffer responsible for arranging a dinner for a group of clients in a city several states away, which will take place after a business conference five months from now. He has many other projects in front of him, with deadlines he must meet in a matter of days and weeks. Instead of conducting the full due diligence on the client dinner to make sure the venue is appropriate and well-regarded and all of the planning logistics are addressed, he makes a hasty decision and picks the first restaurant he comes across in his search. Unfortunately for him, when the dinner actually takes place months later, the clients are aghast at how badly it goes, from the setting to the food to the service, and their impression irrevocably rubs off on the firm, resulting in a loss of business and revenue. All of this transpired because on one busy afternoon, an employee made a quick decision so he could reap the mental reward of clearing it off his desktop. In the long view, early gratification led to a negative impact on his company, and likely on his job. Having said that, we really shouldn't be so hard on the junior staffer. After all, he was reacting in an understandable way to a pressure overload. From the information before him, his brain calculated a series of outcomes and estimated a reasonable chance of success for the dinner months in the offing (and likely a business dinner didn't seem like an especially important project relative to the other duties staring him the face). That calculation redirected resources toward the urgencies of the coming days—a reward in itself, no doubt—and the outcome of the decision was so far in the distance that it could barely compare to the outcomes he would experience if deadlines dropped in the short term. You can imagine a thousand similar examples, with a thousand unfortunate participants, and in each case we would find the decision scale balancing toward immediacy. But short-term thinking is not the only problem, as we'll discuss next.

WHAT I FEEL NOW AND *THINK* I'LL FEEL LATER

"If I had been in that situation, I would have . . ." It's a tired old saw, but all of us have said it at one time or another. And when we say it, we mean it—we are convinced that if we had been in the same situation as someone else, we would have acted differently (i.e., more effectively). My favorite gadfly of this effect was a television show called *What Would You Do?* that placed people in difficult, emotionally charged situations and used hidden cameras to document how they react in the heat of the moment.[3] In one case, an actor plays the role of an employee at a restaurant who hurls racist insults at particular customers (also actors), while other customers look on. Some of the people intervene, but most do not. Instead, they either pretend to ignore the travesty or just watch, dumbstruck and speechless. The host of the show eventually stops the mayhem and tells everyone that they are part of a TV show. He then asks both the people who intervened and the ones who didn't why they acted as they did. It can be uncomfortable to watch those who did nothing explain why they didn't act, and it is all too easy to imagine that we would have been one of the good guys.

What psychology research suggests, however, is that all of us watching that show really have no idea how we would react in a similar situation, unless we have already been in one a lot like it. We falter on something psychologists call the *intensity bias*, which simply means that we are poor forecasters of our emotional reactions.[4] A related bit of psychology jargon is *moral forecasting*—how effective we are at predicting how morally we will act in a given situation. When we are in a non–emotionally charged state (like watching a TV show in the comfort of our living room), we can imagine any number of ways we might react—but all of these future projections are concocted without the presence of emotions we would actually feel in an intense situation. Unless we somehow manufacture a situation to force us into those emotions, there's simply no way we can get there, and certainly not with the intensity we will experience if the actual situation happened (*important sidebar point:* the value of practicing any-thing with as close-to-actual conditions as possible stems from this bias, because if we don't, we are lacking an extremely critical part of what will influence our actions in a real-deal situation).

A similar thing happens when we make a short-term decision without a long-term perspective. In the short term, we are processing a decision that is unplugged from how we will *actually* feel in the long-term. Some-times this works out fine, particularly if the outcome of the decision goes well (the hypothetical employee really could reap major benefits from taking on the new project even though it will be painful to complete). Other times our future forecasting handicap results in problems for us and for others—and though hindsight will tell a different story, we seldom see the problems coming. In addition, we find ourselves in these sorts of situations over and over again, which leaves the impression that for many of us "learning our way out of the problem" isn't possible. It can seem as though we have to repeatedly feel the burn of bad decisions before the point becomes salient enough to change how we think.

And then there is the question of whether we were really paying atten-tion in the first place or floating among the clouds when the decision was being made; we'll address that topic next.

Chapter 4

THE MAGNETISM OF AUTOPILOT

"All that is gold does not glitter; not all those who wander are lost."

—J. R. R. TOLKIEN,
THE FELLOWSHIP OF THE RING

TIME TRAVELING HOMEWARD

You are driving home one night after work, later than typical, with a lot on your mind lingering from a busy day. Most of your thirty-minute drive is on the highway, and as you get on the road, you are pleased to see that traffic is lighter than usual. The tape of the day's tensions and demands replays in your mind. You think back to an especially awkward interaction you had with a coworker who seemed to be accusing you of sabotaging one of his projects. At least, his words struck you that way in the moment. You weren't sure how to respond, so you went with your gut and reacted defensively. With a few hours of reflection, though, you can't help but wonder if perhaps you overreacted. You replay his words, facial expressions, and tone of voice to pick up on anything you may have missed. From his point of view, the situation must have appeared different from how you were seeing it—why else would he accuse you of something so ridiculous? Plus, he's usually a reasonable guy and . . . abruptly, as if you just woke up, you find yourself driving down the street to your house, your driveway now in clear view.

If this scenario sounds, and feels, familiar to you, you're not alone. All of us have felt like we've "lost time" at some point, and can't figure out how we got from here to there without remembering a minute of the transit. When you caught yourself, you probably experienced a jolt of anxiety about what might have gone wrong while you were on autopilot—driving off the side of the road, for example, or running someone over.

Why our brains are happy to switch on autopilot is a topic of intense interest to cognitive scientists. The latest consensus is that most of us are men-tally elsewhere between 30 and 50 percent of our waking hours.[1] Equally striking is the finding that underlies the percentages: Being a space cadet appears to serve an important adaptive function. But like so many adaptive functions, the more we indulge it, the more we are likely to take a fall.

BUILT TO WANDER

The theory of a specific neural structure behind mind wandering is only a decade or so new. Until then, a handful of researchers (most notably Jerome Singer of Yale) suspected that daydreaming was more than a mere bug in the cerebral system and served a useful purpose, but hard evidence to support the theory wasn't available.[2] Brain imaging has helped fill out the picture by showing which brain areas fire up when we're wandering off. Specifically, a web of neurons—dubbed the "default network"—span-ning three brain regions (the medial prefrontal cortex, the posterior cin-gulate cortex, and the parietal cortex) are activated when our brain flips on autopilot.[3] When we're not focused on something capable of holding our interest, the network is triggered. Or, in another view, it's always on in the background but only takes first chair in our brains when we're not focused on anything in particular.

Several theories offer possible explanations as to why such a network exists, but the most compelling, in my view, is that the default network is integral to our sense of self. Imagine a world in which your attention was always focused outward. Your mind would be effectively "externalized," as you would have no opportunity to explore your internal landscape. You would never connect with the "you" within the "you" that interacts with the external world. You would also be unable to process information without focusing on it directly. The default network appears to allow us to digest data as we wander, and most likely also while we sleep. The old adage to sleep on a problem and wake up with an answer is not baseless; in a very real sense, our brains are capable of problem solving in default mode. Quoting the brilliant comedian John Cleese on this point: "When I was in college . . . I would go to bed at night with a problem . . . and when I woke up not only was the solution to the problem immediately apparent to me, but I couldn't even remember what the problem had been the previous night. I couldn't understand why I did not see what the solution was."[4] This is hardly an exaggeration. A solid night's sleep can indeed refashion the elements of a problem in such a way that it's no longer obvious why it was a problem vexing enough to stick around the night before. We may be sleeping, but our brains are doing important work while we're there.

We also know that the default network is triggered by increased stress, boredom, chaotic environments, and fatigue. And, according to research by Harvard psychologist Daniel T. Gilbert, we report being less happy when in mental default, even though (according to this study) our minds are wandering 46 percent of the time.[5] It's difficult to say why we are not especially jubilant in default, but the fact that mind wandering often includes replaying stressful situations probably has a lot to do with it. Whether or not we assess ourselves as happy in autopilot, what is clear is that our brains are "happy" to go there; all other things being equal, our noggins will drift ethereal without fail. It's more than a tendency; it's a well-worn biological pathway.

On the upside, research points to a strong link between mind wandering and creativity.[6] This seems especially true for those of us who can pull ourselves out of our daydreams at will. The ability to retreat into the clouds and come back down to terra firma may be one of the more effective self-preserving functions we possess, enabling us to extract ourselves from unhealthy environments and relocate in mental space of our choosing. In that space, we are less constrained to envision creative solutions to problems or simply allow ourselves to paint boundless images across our mental canvass. The philosopher Bertrand Russell touches on this point in his book *The Conquest of Happiness* when he said:

> The man who can forget his worries by means of a genuine interest in say . . . the history of stars, will find that, when he returns from his excursion into the impersonal world, he has acquired a poise and calm that enable him to deal with his worries in the best way.[7]

Happier When Busy, Wired to Be Lazy

If you ever watched the show *Fraggle Rock* from the eighties, you'll remember that the Doozers were little creatures who spent all of their time building things. Unfortunately for them, the Fraggles—a far lazier critter—loved to eat the Doozers' buildings (though not the Doozers themselves) and summarily crushed the product of the little creatures' hard work anytime they wanted a snack. But the Doozers never seemed the least bit frustrated by this and just kept right on building. Psychology research suggests that we're better off being like the Doozers, though we're wired more like the Fraggles. In one study, participants were offered an identical reward (a chocolate candy bar) for either delivering a completed questionnaire to a location that was a fifteen-minute walk away, or delivering it just outside the room they were in and then waiting fifteen minutes. Sixty-eight percent chose to deliver it just outside the room and wait. When the reward was changed to a slightly different chocolate candy bar, 59 percent chose to walk fifteen minutes to deliver the questionnaire (and this held true even though both types of candy bars were rated as equally appealing by all participants). Afterward, participants who took the walk rated themselves as feeling significantly happier than those who sat it out. It appears that our first instinct is for idleness, but when given an excuse to be busy (even a meaningless one), we're liable to act on it and consequently feel happier. But before you go looking for busywork, remember that our evolutionary vestige to conserve energy is tough to overcome. Believe it or not, laziness, in marginal doses, serves a purpose.[8]

LOST IN DEFAULT

Overindulgence of our brain's tendency to wander is, however, a potentially debilitating problem. Psychologists have even coined a phrase to describe it: *obsessive rumination*. Those of us who obsessively ruminate are prone to lose ourselves in dark otherworlds of our creation. Those worlds and the freedom they provide have a compulsive quality not unlike the escapism of role-playing (which we'll be discussing in the next chapter). What also seems true is that rumination comes in more than one flavor, some more directed than others. While you may be able to more or less manage the flow of thought as your mind glides, I might find myself drifting aimlessly. The greater the degree of direction one is able to exert on the mind's wanderings, the greater one's ability to pull out from the drift and come back to the here and now as Russell describes. That is much harder to do than it sounds, chiefly because the default network's activity is strongest when we don't know we are drifting. That was the finding of researchers from the University of British Columbia studying the neural trip wires that jettison our attention.[9] The deeper and more complex our drift into the ether becomes, the more our mind-space is consumed.

Worth noting: Something very similar happens when we drink alcohol. Have you ever had a couple of drinks and your mind started wandering in a booze-induced haze? If you can't remember that happening, that's probably because you had a couple of drinks. Research suggests that alcohol has the dual effect of causing our minds to wander while not noticing that we have zoned out, which is essentially an amplified version of our normal tendency to drift (and, as we've seen, our brains really don't need any help going there).[10]

Research indicates that those who obsessively ruminate tend to dwell on negative thoughts and emotions. There is, in fact, a strong correlation between this kind of rumination and depression. Repeating mistakes, hurtful remarks, stressful situations, and the like over and over again in the mind is like being stuck in a self-defeating movie of our own making, starring us. And when the movie starts, it is hard to turn off largely because our penchant for the drift—even when filled with dark thoughts—is built into our brains.

We have the capacity to pull ourselves out of the drift—to stem rumination—and it comes down to exerting "mindfulness," or what we can more specifically call *metacognitive awareness*, our ability to think about our thinking (a topic I address in detail in my book *Brain Changer*). A great deal of research on mindfulness and metacognition has been carried out in the last several years, with a consistent outcome: we can achieve leverage over the direction and intensity of the drift, but it takes effort and practice.

THE ATTENTION GAME

The topics discussed in this chapter naturally align with another, broader topic: how we direct our attention. As the "attention economy" becomes ever more aggressive, the need to effectively manage how we direct our attention becomes more important. A few things are worth noting here. First, just as overindulgence in default mind wandering can lead to rumination, overindulge in distraction (of our active attention) can take a cumulative toll. And the truth is much of the time we don't even realize we're that distracted. Much of what chips away at our attention are little things dancing on the fringes of awareness, fragmenting our focus moment by moment.

The science of attention tells us that our brains aren't too bad at filtering out big distractions. We can usually regulate our reactions to in-your-face attention grabbers. The first few billboards on the highway might get you to look, but a couple of minutes later you're driving along like they're not even there. When a distraction is overt, the brain adapts. But when distractions are subtle, it's a different story.

For Your Brain, Downtime Is Never Just Downtime

Your brain is *always* doing something. Even when you think it's "offline" because you aren't actively engaging a task or problem, your brain is busy reducing activity in some of its areas and increasing it in others. The question that has preoccupied neuroscientists is, what's the goal of all of this brain activity when we aren't outwardly doing anything in particular? An intriguing study offers an answer that makes Facebook browsing suddenly seem more meaningful. The researchers believed that when the brain isn't actively engaging in a task, it drops into a mode that prepares it to be social with other brains. "When I want to take a break from work, the brain network that comes on is the same network we use when we're looking through our Facebook timeline and seeing what our friends are up to," says Dr. Mathew Lieberman, study coauthor from the University of California, Los Angeles.[11] "That's what our brain wants to do, especially when we take a break from work that requires other brain networks."

The research team observed participants' brain activity using fMRI while they made a series of fast "yes or no" judgments under a few different conditions, some involving social elements and others not. The researchers found that they were able to make predictions about how participants made social judgments by gauging activity in a brain area called the dorsomedial prefrontal cortex (DPC). As it turns out, the DPC is also central to the brain's "default mode"—the mode we drop into when we aren't actively engaging in tasks.

As we discussed, daydreaming is important default-mode time for the brain during which less directly-active processing happens. This study suggests that part of this default-mode processing helps the brain prepare for social interaction with other brains. According to Lieberman, "The brain has a major system that seems predisposed to get us ready to be social in our spare moments. The social nature of our brains is biologically based."

The DPC also happens to be an energy-intensive brain area—it takes a lot of circulating blood glucose to fuel our allegedly non-active brain processing time. Knowing that should inspire new respect for what our brains are accomplishing while we think we're doing nothing. Downtime is never simply downtime—it's just a different sort of processing time for the perpetually processing brain.

One study tripped on that discovery by accident. Researchers were trying to identify which sorts of stimuli attract the most attention (with the hypothesis that the most rewarding stimuli—in other words, things with a payoff—have the greatest draw); unexpectedly, they found that subtle, under-the-radar distractions trigger the biggest reactions.[12]

Jeff Moher, one of the study's authors and a visiting assistant professor of psychology at Williams College, told me that the finding took the research team by surprise:

> We initially expected to find that highly rewarding stimuli automatically attract action responses. We got some pilot data with the exact opposite effect. This caught us by surprise, so we took a step back to see whether our assumptions about the kinds of objects that are distracting held true for goal-directed action responses. What we found . . . is that our assumptions were wrong—in fact, the more salient distractor produced less action interference than the less salient distractor.[13]

In other words, the smallest distractions triggered the biggest reactions.

Moher explained that he thinks this is largely about "suppression"— the brain's ability to quickly identify and snuff out certain distractions.

> One explanation we think is reasonable is the idea that suppression is triggered only by a sufficiently strong distractor. In other words, maybe an object that really jumps out from its surroundings but is not relevant to the task at hand can be relatively easily ignored. An object that only slightly jumps out from its surroundings might be enough to disrupt your ability to move to the target, but not so strong as to trigger this suppression.[14]

That makes sense when you consider that suppression is an energy-intensive brain activity. It takes a lot of cerebral juice to block out distractions, so the brain—an energy miser at its core—reserves this expenditure for the big stuff. Consequently, a fair amount of small stuff slips through the filter.

Of course, this isn't an absolute rule by any stretch. Plenty of big things are capable of triggering an immediate action response, too, but there's generally a compelling reason for the response (or at least one we're trained to perceive as compelling). It would be hard to ignore a blaring fire alarm even if you knew you didn't have to evacuate.

These findings serve as a useful backdrop to those of another study showing that the ongoing, subtle notifications on your smartphone are fragmenting your attention more than any larger purpose the phone serves. Researchers leading the study found that any sort of notification— a sound, a buzz, a vibration—triggers "task irrelevant thoughts" (another term for mind wandering), even though we may not immediately pick up the phone.[15]

Again, it's the small stuff. And in our distraction-saturated world, the latest science tells us that it's exactly that small stuff our brains need the most help blocking out.

Moreover, as we know, the "small stuff" can also be a matter of life and death. Choosing to text while driving ends more lives every year. To put a finer point on that: *failing to manage our attention* when doing something as risk-intensive as driving is ending more lives every year. Each text seems like a small thing, but only one small slip of attention on the road can have a catastrophic result.

Managing attention is no small matter, no matter how small the distraction.

Chapter 5

IMMERSION AND THE GREAT ESCAPE

"Beware lest you lose the substance by grasping at the shadow."

—AESOP, *AESOP'S FABLES*,
"THE DOG AND THE SHADOW"

WHEN THE WEB WAS YOUNG

I distinctly remember the day I first saw the World Wide Web (WWW). It's one of those "where were you when" moments that usually append a historic tragedy, but in this case it was the unveiling of a technology that would change the world in ways both fantastic and tragic.

It was 1993, and I was in the University of Florida Interactive Media Lab, where a small group of classmates and I convened to witness what at the time was described as a "graphical overlay for the World Wide Web," called Mosaic. Only a minority of people knew such a thing as the WWW existed at all. In its monochrome, all-text form, there was little in it to excite the masses. Information hounds and users of bulletin board systems (think of 2400-baud modems) had patronized the network for some time, but for the world to take notice, something bigger had to happen—something expansive well beyond the interests of techno-savvy enclaves. Well, big happened, and then some.

We were floored, speechless, almost unbelieving as the images scrolled before our eyes. As students of media technology absorbed in the history of mediation and the future potential of the new media, all of us knew we were witnessing the microexplosion of a technological big bang.

Less evident at the time was how this new universe would affect entire populations—indeed, *successive* generations of populations—across the physical world. How could anyone have known? The chronicles of media, while littered with history-altering shifts, had little to say about the effects of a deeply immersive technology, particularly one with powers of organic growth rivaling the speed and potency of thought itself.

A multitude of predictions were made, of course, and this chapter (and entire book) couldn't hold them all. For our purposes, one matters more than the rest, though the decades since that day in 1993 have proven it an inadequate forecast at best.

But before we discuss immersive e-media and its consequences, I want to zip back in time a couple of decades before Mosaic shook the world, before technologically enabled escapism was the only game in town. These were the days of Dungeons & Dragons, played out of books (actual, paper books), with dice, pencils, and paper. Arguably the most immersive gaming experience ever invented up to that point, D&D, as it was called, was to outsiders frightening, bordering on dangerous. To some, it was simply diabolical. Every few months, a news story about a D&D–playing teenager jumping off a building or stabbing another teenager implicated the game as the insidious influencer. Christian television was brimming with scathing sermons about the game Lucifer himself must have crafted.

Most of these reactions—religious and sectarian—were overblown rants. D&D was the target de jour for anyone with a pulpit or soapbox, and more than a few of the criticisms against it were little more than silly. But what was plainly true is that D&D did enable something undeniably powerful: an immersive experience with a rabidly compulsive quality.

The formula for successful role-playing, in any form, is to empower participants to assume an identity as far removed from their day-to-day identities as they wish, and provide a dynamic, interactive world for them to explore. While the parameters of physical existence are all too apparent, the new identity is beginning a virtually limitless existence. The appeal of living, in effect, another life, one that consumes the same cranial space as the one that requires breathing air and drinking water, cannot be understated. And it should be noted that each of us comes equipped to "spin off" identities separate from the one consuming most of our waking hours, and these identities find a way to emerge, whether via role-playing or other experiences that fit the need.[1] For example, the "you" who curses other drivers up and down the highway is probably not the "you" who behaves diplomatically at the office. In this sense, taking on new identities via media immersion in any form is not an extraordinary stretch for our brains; it just happens that some forms of immersion are more compelling than others.

Despite what now seem like anachronistic tools, D&D was to role-playing what the Macintosh was to desktop publishing. Players (and in my early teens I was one of them) invested mental energy in the game at a level unappreciated by anyone who had not personally become immersed in the fantastic worlds that were instantly accessible when a new game module was opened. With years to digest and reflect on the experience, what playing D&D taught me most is that our brains are quite capable of participating in split existences—at least to a point.

It is this "point" that deserves the most attention, because immersive e-media have blurred the distinction between "existences" beyond what anyone might have predicted in the early days of role-playing. It would have been hard to imagine, for example, thousands of Internet cafés strewn across Southeast Asia, packed with patrons twenty-four hours a day. And few would have looked down the road and seen parents abandon their children in favor of virtual children whom they nurture online day and night while their real children die of malnutrition and dehydration in the same room. Could anyone have envisioned a problem big enough that a government would issue laws that force establishments with Internet access to close up shop at night to prevent online junkies from never leaving? These are not fictional examples. Over the last few years, several countries, including South Korea and Vietnam, have been struggling with how to handle online role-playing "addiction," to little avail.[2]

When it comes to e-media immersion, we walk an extremely fine line. Unfortunately for us—some more than others—our brains lean toward the wrong side of that line, and the consequences can be severe. And so it is with many other activities that seem to navigate their way into our brains so quickly that we fail to notice that it's happening. Why does our brain seem to want more of what ultimately isn't best for us?

PERILOUS STROLLS DOWN THE COMPULSION PATHWAY

To adequately tackle the issue of compulsive e-media use, or any compulsive behavior, we need to take a few minutes and dive into what cognitive-science research tells us about why our brains are prone to these behaviors to begin with. Right up front, a qualifier to this discussion is in order. Quoting clinical psychologist and psychoanalyst Todd Essig: "While all addictions become compulsive, not all compulsions are addictions."[3] This is a crucial point, because it is too easy to put all compulsive activity into the same oversimplified bucket and ignore the differences—something popular media does habitually.

Again quoting Dr. Essig: "Nonaddictive compulsions can really screw up a life as much as can being a junkie: hand washing, walking backwards, calorie restrictions, body modifications, facial tics, plastic surgery, etc."[4] Whether sleepless Korean denizens of video-game parlors are more like junkies or like people with OCD or Tourette's—or are victims of habit reinforcement run amok—is still very much an open question. So the distinction between claiming someone is "addicted" to immersive media, or is "compulsively drawn" to it, is one worth acknowledging.

What we know is that our brains are equipped with a "reward center" that serves to adaptively motivate behaviors that benefit our species. These behaviors include every rudiment of survival: nourishment, care of the young, and sex, to name a critical few. And they include behaviors that help us thrive in our environment—what we can loosely describe as achievement motivators. Without this motivation and drive to seek out pleasurable experience, we would be a very dreary and endangered species. We depend on a well-functioning reward center to drive us toward goals, from the most minor to those that will change our lives. Each goal, no matter the topic, is to our brains a reward. As noted earlier in the book, our brains have evolved around capacities for pattern-seeking and recognition, and reward identification and pursuit. Boil it all down and that's what we are: pattern finders and reward seekers. And the reward center is the apparatus that enables one of those central capacities.

This reward center (called the *mesolimbic reward center*), while indispensable to us, is not unlike an unprotected power grid in that it can be hijacked and tapped into from external forces. These forces make use of the same reward circuitry that benefits us in so many ways, and this circuitry (called *incentive salience circuitry* or ISC) adaptively responds just as it does to accommodate beneficial rewards. The problem is that the new rewards imprinting the ISC are generally not beneficial. But our brains suffer a sort of reward-distinction blindness, and new imprints are integrated into the grid.[5]

When the reward center is overwhelmed by disadvantageous reward imprinting, we say that someone is "addicted" to a substance or behavior. At this point, the reward center is in a state of malfunction, a tenaciously difficult problem to reverse. As the neural underpinnings of addiction have been better understood, so has the understanding that the common denominator in all compulsive behaviors is a malfunctioning reward center. Whether drug abuse, gambling, overeating, or sexual behavior is the culprit, the same underlying dynamic facilitates compulsive continuation and intensification of the behavior.

Early neuroscience research on addiction used our friends the rats to test the theory that stimulation of the reward center can become compulsive. Tiny electrodes were inserted into the part of the rats' brains thought to be responsible for motivating pleasure-seeking behavior. The rats were then trained to press a bar that activated the electrodes, giving them a "hit" to the reward center. Before too long, it became apparent that the rats more than enjoyed the hit, because they wouldn't stop pressing the bar. In fact, they would not eat, drink, sleep, or have sex as long as the bar was available. Many collapsed from exhaustion, and the rats not forced to eat starved themselves to death—but they never gave up the bar.[6]

This explains much about why meth and crack addicts are willing to forgo food, sleep, and sex to get more of the substance that their brain now craves the most. And it also suggests an explanation as to why parents would abrogate responsibility to nurture their own child so that they could continue receiving the overpowering reward of staying immersed online. Though the processes at play are not identical to those resulting from chemical addiction (chemical addiction is more acute and overwhelming to the brain's circuitry), it is clear that e-immersion is providing a reward, and that continuing to seek the reward (in this case, continuing to "raise" a mystical child online) has become compulsive. The more the reward is sought, the more the compulsive behavior is reinforced.

The neurochemical currency at play in these and all addiction scenarios is dopamine; specifically, dopamine receptor activity in a part of the brain called the *ventral tegmental area* (VTA). Often called the "reward neurotransmitter," dopamine is essential to our survival, but a potent enemy within when our brain's reward circuitry is overwhelmed with the wrong kinds of rewards. (See Special Section 2 for more about the technology connection to the reward center.)

Clearly some people are more susceptible to compulsive behavior than others; a genetic component cannot be denied. But what's alarming is that anyone's brain can theoretically become addicted to a substance or behavior, given enough exposure. And once that happens, the addiction pathways are open to accommodate additional compulsive behaviors, which is why addictions to narcotics and sexual behavior or gambling often present in the same person. Said another way, the brain's incredible plasticity—it's uncanny ability to adaptively change—can ensnare us from the inside out.

WHICH BRINGS US BACK TO LIFE ONLINE

The reason I am choosing to focus primarily on immersive e-media in this chapter, and less on the droves of other compulsive behaviors any of us could trip on, is simply because it is ubiquitous—indeed, ubiquitous at an unprecedented level—and will only become more so. We have seen only the beginning of the expanding online universe. As we become ever more connected to our smartphones, the effects explored in this chapter are only increasing and strengthening. Added to that, research indicates that immersive e-media are an effective *compulsivity accelerant.*

In his book *iBrain*, Dr. Gary Small, director of the UCLA Memory and Aging Research Center at the Semel Institute for Neuroscience and Human Behavior, explains that particularly for those bringing compulsive tendencies to the technology, the results can be severe. Quoting Dr. Small, "Someone with obsessive-compulsive tendencies is already predisposed to a range of addictive behaviors, and technology has a way of accelerating that process."[7] This is an important point because it is unclear what percentage of the population is predisposed to compulsivity disorders, although some estimates put the numbers as high as fifty million people in the United States alone.[8] Even half that estimate represents a significant percentage of the population in just one country, to say nothing of the developed world overall.

One reason for this hypercompulsivity effect may be the neural rewards gained from the sense of belonging many of us get from online interaction. Scott Caplan, associate professor in the Department of Communication at the University of Delaware, commented on this subject:

"The research suggests that people who prefer online social interaction over face-to-face interaction also score higher on measures of compulsive Internet use and using the Internet to alter their moods."[9] In 2007, Caplan conducted a study of 343 undergraduate students to determine the major variables contributing to problematic Internet use (or PIU, to use the standard term). He examined personality-related variables, such as loneliness and social anxiety, but also potentially compulsive activities like playing video games online, viewing pornography, and Internet gambling to determine which stoked the fires of compulsive use the most.[10]

Of these variables, social anxiety emerged as the strongest. "The argument I've made in my research is that individuals who have problems with face-to-face interactions are drawn to the unique features of being online," Caplan adds. "As they develop a preference for online social interaction, my hypothesis is that they begin to use these channels for mood regulation which becomes compulsive."[11]

Admittedly, it is important not to paint everyone with the same digital brush. Just as with exposure to alcohol or gambling, some people are bitten and begin the compulsive drive immediately. Others seem unaffected even after deep exposure. This is undoubtedly also true of life online. One crucial difference, however, is the extent to which larger swathes in nearly every age group of the population are exposed to immersive e-media versus most other potentially compulsive behaviors. As I write, it is yet unclear what the long-term effects of such broad exposure will be. What is clear is that there is enough research-based evidence to warrant a cautious—though not paranoid—approach to the future of life online.

THE BRAIN THAT WANTS TO BELONG

Some of this evidence comes from studying the former heavyweight champion of e-media: television. Since the 1970s, psychological research of television's effects has scarcely abated, even as the Internet began swallowing larger and larger portions of our attention. As it turns out, television has admirable staying power as an attention grabber, despite how many other diversions emerge. And because it has been with us for so long (many decades longer than the online world), it is still one of our best sources for finding out how e-media affects our brains.

One research team engaged the topic by asking whether we can become emotionally entangled with fictional characters on a TV screen. Four studies conducted by researchers at the University of Buffalo tested the "social surrogacy hypothesis," which holds that humans can use technologies, like television, to feel a sense of belonging that they're lacking in their physical lives.[12] And not only TV, but movies, music, and video games can fill this need as well, according to the theory. The experiments measured different categories of emotional reaction, like self-esteem, belongingness, loneliness, rejection, and exclusion, in response to descriptions of subjects' favorite TV programs. In one of the experiments, 222 undergraduate students were asked to write a ten-minute essay about their favorite TV shows, and then to write about TV programs they watch when nothing else is on, or about the experience of achieving something noteworthy in school. Afterward, they were asked to verbally describe what they had written in as much detail as possible. The results: After writing about their favorite TV shows, participants verbally expressed fewer feelings of loneliness and exclusion than when describing the filler TV shows or the experience of academic achievement. The takeaway is that surrogate relationships with characters or personalities in TV programs can fill emotional needs. Another of the experiments produced results suggesting that thinking about a favorite TV show buffers against drops in self-esteem and feelings of rejection that accompany the end of a relationship—an electronic vaccine for heartbreak. These results buttress a psychological concept that most people would admit scares them a little: "technology induced belongingness." That spending a half hour with our favorite imaginary personalities can turn on our love lights seems a bit strange but may be truer than we're willing to admit.

The Neural Connection between Loneliness and Conflict

Social neuroscientist John Cacioppo conducted a 2009 brain imaging study to identify differences in the neural mechanisms of lonely and nonlonely people. Specifically, he wanted to know what's going on in the brains of individuals with an acute sense of "social isolation"—a key ingredient in loneliness that has nothing to do with being physically alone and everything to do with *feeling* alone. While in an MRI machine, subjects viewed a series of images, some with positive connotations, such as happy people doing fun things, and others with negative associations, such as scenes of human conflict. As the two groups watched pleasant imagery, the area of the brain that recognizes rewards showed a significantly greater response in nonlonely people than in lonely people. Similarly, the visual cortex of lonely subjects responded much more strongly to unpleasant images of people than to unpleasant images of objects—suggesting that the attention of lonely people is especially drawn to human conflict. Nonlonely subjects showed no such difference. In short, people with an acute sense of social isolation appear to have a reduced response to things that make most people happy, and a heightened response to human conflict. This explains a lot about people who not only wallow in unhappiness but also seem obsessed with the emotional "drama" of others.[13]

REALTY AND FICTION ARE ONLY PIXELS APART

Results like those from the study just described spark a crucial question: How does the brain distinguish between reality and fiction—indeed, *does* the brain distinguish between reality and fiction?

This question served as the jumping off point for another study, conducted at the Max Planck Institute for Human Brain and Cognitive Sciences, which attempted to identify how the brain responds when exposed to contexts involving real people or fictional characters.[14] The research followed up on a similar study conducted in 2008 titled "Meeting

George Bush versus Meeting Cinderella: The Neural Response When Telling Apart What Is Real from What Is Fictional in the Context of Our Reality."

In the present study, researchers used functional magnetic resonance imaging (fMRI) to evaluate subjects' brain regions—specifically the anterior medial prefrontal and posterior cingulate cortices (amPFC, PCC)—while they were exposed to contexts involving three groups: (1) family and friends (identified as a high-relevance group), (2) famous people (medium-relevance), and (3) fictional characters (low-relevance). The working hypothesis was that exposure to contexts with a higher degree of relevance would result in stronger activation of the amPFC and PCC.

In previous studies, the amPFC and PCC were shown to play a large role in self-referential thinking and autobiographical memory (essentially, the "me" part of your brain). The idea behind the present hypothesis is that information about real people, as opposed to fictional characters, is coded in the brain in such a way that it elicits a self-referential and autobiographical "me" response. The more personally relevant the context is, the stronger the response. The results were consistent with the hypothesis, showing a pattern of activation in which higher-relevance subjects were associated with stronger amPFC and PCC responses. This result also held true for several other brain regions to varying degrees. In other words, for our brains, reality equals relevance. But then, how do we end up merging with fictional (or virtual) personalities on TV or a computer screen? For the answer we have to delve into what "relevance" really means in our day-to-day lives. The truth is many of us spend more time connecting with personalities online and on TV than we do with people physically in our presence.

Social neuroscientist John Cacioppo, whose research on loneliness has opened new doors to understanding this darkest of emotions, has observed that we live in an age of constant interaction, and yet more of us are claiming we're "lonely" than ever before. Loneliness, Cacioppo points out, has nothing to do with how many people are physically around us, but has everything to do with our failure to get what we need from our relationships, which manifests as "feelings of loneliness" regardless of how many names and faces are in our social orbits.[15] The research tells us that virtual personalities online and characters on television are surrogates for emotional-need fulfillment, and hence occupy the blurry margins in which our brains have difficulty distinguishing real from unreal. The more we rely on these personalities and characters for a sense of "connect-edness," the more our brains encode them as "relevant."

All of this can be said another way: Our brains can be tricked, and the irony is that we're complicit in the trickery. Though it's arguably a stretch to call it "trickery" if you consider these effects in the context of how the brain is natively structured to identify and pursue rewards. As reward-driven animals, we seek out the paths of least resistance to get what we need (rewards), in whatever form they take, and electronic immersion provides the most accessible, nonchemical reward path yet invented. It's not so much a matter of trickery as it is our brains simply doing what they've evolved to do, and technology providing accessible paths to do it.

PARTING THOUGHTS

The ongoing debate about media effects, flanked by extremes on both sides, is always a couple of inches away from getting out of hand. Media in any form does not, strictly speaking, *cause* any human behavior. To suggest otherwise is to endorse the old "blank slate" notion of human nature that was abandoned a long time ago. We bring a complex array of psychological variables to the media experience—as we do with all experiences—and it's the interplay between what we bring and what we see, hear, and read that yields influences on our thinking and behavior. For some, these influences will be subtle and virtually unnoticeable. For others, they can be significantly more pronounced and damaging.

We also have to remember that people *choose* to immerse themselves in any form of media—or, truly, anything else. While the reward system is open to hijacking by external forces, there is always an original motivation for deciding to do what we do, and there are always decision points along the way where we choose to participate or withdraw.

A full explanation of why people go there in the first place is well beyond the scope of this book. The big takeaway is that we are living in a world with ever more compulsive diversions, and that is only going to intensify in the years ahead. We are all carrying around extremely powerful diversion devices in our pockets and purses, with all signs pointing to them getting only more powerful with time. Careful consideration of the outcomes from our brains interacting with these immersion technologies of our making—as opposed to paranoia or blanket indulgence—is what's needed now. (For more on the reward center and its relationship with technology, check out Special Section 2.)

Part 3

MOTIVATION, RESTRAINT, AND REGRET

Chapter 6

REVVING YOUR ENGINE IN IDLE

"Thunder is good, thunder is impressive; but it is lightning that does the work."
—MARK TWAIN, FROM A LETTER
TO AN UNIDENTIFIED PERSON

BEATING THE SYSTEM LIKE A DRUM

I'd like to introduce you to a most unlikely master of industry. We begin by observing Mike in high school in the late 1980s. He is, to use the vernacular of the eighties, a "total burnout." He misses classes and chronically shows up late for the ones he does attend. The only thing he makes sure never to miss are parties. At those, he's a fixture. He sleeps very little and parties very hard—and he's a thread shy of being kicked out of school.

What almost no one realizes about this young man is that he really does have a quite lofty aspiration: He wants to learn everything there is to know about lasers. From a very young age, he was fascinated with light—from flashlights to fluorescents—and was endlessly curious about how they worked. School bores him. His real education, as far as he's concerned, comes from reading what he wants to read. Sometimes he spends hours at the local library, reading books about laser technology—how it works, its applications, its future. It's the one topic that engages him, and he tackles it full force.

The other thing just a few people know about him is that he's a textbook lazy genius—an underachiever par excellence. While he struggles to get Cs in school, he aces standardized tests without the least preparation. And that's how he gets to college, where, with more freedom to learn what he wants to learn, he does exceptionally well. So well, in fact, that he's offered a spot in a laser tech graduate program at Carnegie Mellon University. He studies with some of the most renowned laser experts in the world. While still in school, he starts his first business developing specialized lasers for computer hardware companies. He eventually starts three more companies, each focused on development of customized laser applications.

Today, he is one of the foremost laser experts in the world, and the pioneer of pathbreaking laser applications most of us can barely imagine. His companies work with the largest corporations on the planet; in a very real sense, he is helping to design the future.

Mike's story is an illustration that the systems we find ourselves in are not always best suited for our potential. Low achievers, even extremely smart ones, can become terminal failures if their passions and interests are not tapped. Educational systems accomplish this in some cases, but often they simply fail. That leaves people like Mike to either figure out how to work around the shortcomings of the system themselves, or be left in the dust. Mike did it by focusing on the one thing that was always fun for him to learn about from an early age.

To make matters more complicated, the happy brain is not natively structured to challenge the systems we inhabit. Think of a system—educational or otherwise—as an environment that has been built for people like us. Schools are, in fact, just that: designed environments for human learning. Once we become part of that environment, our brains begin the work of mapping out the territory so that we can secure a niche. When that has been achieved, changing things up causes instability, and instability is a threat to the happy brain.

So how does a chronic low achiever like Mike (or one of the other many examples out in the world) go from nearly not making it through high school to becoming a global master of industry? That question brings us to our first topic.

ACHIEVEMENT FOR ME, BORING FOR YOU

With Mike's story as a backdrop, let's start with a question about what sort of achiever you are. If I asked you to place yourself somewhere along a spectrum, with the far left side representing quintessential slackerdom, and the far right side representing hyper-overachiever mania—where would you fall? We'll come back to that question in a minute, but first let's talk about an age-old generalization.

Everyone agrees that high achievers do many things well, particularly when they're convinced that excellence requires their utmost performance. Low achievers, as we also know, have a hard time getting motivated and often find themselves coughing in the dust of the high achievers' hustle. That's the generalization, and, like all generalizations, this one has a definite limit. A study conducted by University of Florida researcher William Hart uncovered a variable that knocks this scenario on its head, and it has everything to do with what makes low achievers tick.[1]

Researchers conducted multiple studies to evaluate how participants' attitudes toward achievement influenced their performance. In one study, participants were primed with high-achievement words (related to winning, excellence, etc.) flashed on a computer screen. Each word appeared only for an instant, too fast for conscious deliberation. Participants with high-achievement motivation performed significantly better on tasks after being primed with the words than those with low-achievement motivation.

In another study, participants completing word-search puzzles were interrupted, and then given a choice to either resume the task or switch to a task they perceived as more enjoyable. Those with high-achievement motivation were significantly more likely to return to the puzzles than the underachievers.

The results of those studies buttress what we generally know about high and low achievers. But the final study was a wicked curveball. Participants were primed with high-achievement words (e.g., *excel, compete, win*) and then asked to complete a word-search puzzle. But instead of describing the task as a serious test of verbal proficiency, the researchers called it "fun." The results: Participants with high-achievement motivation did significantly worse on the task than low achievers.

The study authors believe that when high achievers are primed to achieve excellence, the idea that a task is "fun" undercuts their desire to excel. If something is enjoyable and fun, how could it possibly be a credible gauge of achievement? Conversely, low achievers who are similarly primed with achievement words perceive a "fun" task as worthwhile. Not only is their motivation to perform improved, so is their ability.

This intriguing twist says much about why one-size-fits-all educational strategies so often fail. For students motivated to achieve excellence, making tasks entertaining may actually undermine their performance. Likewise, for those not normally motivated to achieve, describing a task as urgent and serious yields the predictable result.

Coming back to my initial question: Whether your answer put you closer to the low-achiever or high-achiever end of the spectrum, the takeaway is that trying to force yourself into a motivational mold not sized for your personality probably isn't going to work. Regrettably, many of the "systems" we find ourselves in—educational or otherwise—were designed without this knowledge, so it's usually up to us to figure out how to game our way into better performance.

The crucial thing to remember is that your brain is not natively tuned to challenge the system. Going head-to-head with established conventions causes disruptions in stability and consistency—and that triggers alarms. You can heed those alarms and stay right where you are, or you can forge ahead and find a way to manage the conflict. For example, you may be the sort of person who finds getting motivated at work extremely difficult. You can fight through your lack of motivation as a matter of necessity, but it's a draining process that just doesn't seem like it should be so hard. It's time to more closely examine what sort of achiever you are and determine if the motivational dynamics you rely on are the best fit. If you think you're closer to the low-achiever end of the spectrum, ask if you have enough enjoyment threaded through your projects.

The remedy might be very simple, like listening to music while working. Or it might be more elaborate, like scheduling breaks in your planner to get out of the office at different points during the day and talking with a colleague about movies, music, or anything else that will inject a dose of fun into your day. Or you may need to audit your projects to identify ways to "lighten" your disposition toward their more mundane requirements. Push yourself to be creative in finding the solution.

If you find yourself always well behind projects and chores at home, change your perspective and ask if you're having a hard time getting things done because you haven't engaged the right motivational cues for who *you* are. If you like gardening because it's a fun, laid-back pastime, but hate organizing your garage (for obvious reasons), do a little fun-infused self-negotiation. On Saturday, agree to devote three hours to organizing the garage. On Sunday, you're not even going to think about the garage, but instead spend as much time cultivating your garden as you like. Spending time doing what you enjoy becomes, in effect, a reward for accomplishing what you don't.

It's easy to get caught up in the popular mantra of achievement that simplistically assumes motivation is always available as long as the achiever has enough desire and will. The point is that lacking a hard-wired desire to "achieve" does not mean you *can't* achieve, or even that you're "achievement handicapped." The research indicates that you just need to be a little more creative. In doing so, the fight to reach your goals becomes less arduous and might even become something you like taking on. Remember, your brain is structured to take the path of least resistance, because that's the less threatening and less stability-sacrificing way to go—but it's usually not the one that will lead you to greater echelons of achievement.

COMPETITION AND THE SIZE OF YOUR FISHBOWL

Now let's shift to another motivational vantage point and examine what sort of competitor you are. Are you a "solo practitioner" or do you perform best when you can see your competition, sizing them up before battle? If you think you're in the second category, then you're just like most of us. Conventional wisdom has it that one of our mightiest competitive motivators is social comparison: We begin competing with others as soon as we compare ourselves to them. Whether the stakes are minuscule or massive, something in us wants to measure up inch for inch. Research conducted by University of Michigan professor of psychology Stephen Garcia shows, however, that our competitive urges don't engorge in a vacuum: It's not merely being among a group of competitors, but the number of competitors we're vying against that has a direct effect on our motivation to compete.[2] Your brain isn't about to let you dive into competition without opening an opportunity for you to short-circuit your efforts and get back to a comfortable, noncompetitive state of mind.

Here's an illustration: Jessica takes a seat in a classroom with ten other students. She looks around, evaluates the competitive landscape, and determines that her odds of doing well against this small group are good. The instructor passes out the particle physics exam, and Jessica is off and running, motivated to score among the best in this class.

Jason arrives at a different room to take his exam, and it's a lot bigger than Jessica's. In fact, it's ten times as big, and Jason has to find a seat in a crowd of one hundred students. He looks around and gulps. There's no way to realistically compare himself to so many people. The instructor passes out the exam and Jason begins without feeling a competitive edge.

The lack of motivation that Jason feels, in comparison to Jessica's hypermotivated resolve, is what psychologists refer to as the *N-Effect*: the effect that occurs when the number of total competitors results in diminished motivation for individual competitors. Researchers assessed this effect through a series of five studies: the first examined SAT and CRT (Cognitive Reflective Test) scores in light of how many people took the tests in given venues over multiple years. Even when controlling for other variables, researchers found a significant inverse correlation between the number of test takers and scores: the more people taking the test, the worse the scores. Another study examined whether test takers, told to finish the test as quickly as possible, would finish their test faster when competing against ten others or when competing against one hundred. As predicted, the best-scoring testers finished their tests significantly faster when competing against a smaller group.

WHY SELF-AWARENESS FUELS MOTIVATION

What's the best way to keep the N-Effect from undermining your motivation to compete? As with many unconscious influences, the solution is to identify it early and critically dismantle its effect before you succumb. In other words, force yourself to exert more rational muscle than you would if blind to the influence.

For example, let's say that you are interviewing for a job, and as you enter the office lobby you see six other candidates awaiting interviews for the same position. As you take your seat, you think to yourself that if six others are there *that* day for the interview, then it's more than likely several others are interviewing as well—making the total candidate pool far larger than you had expected. Your first reaction is that your chances of getting this job just took a harrowing drop. That thought leads to a jolt of intimidation, and your motivation is falling by the second.

But you stop yourself right there and ask, "If I didn't know how many other candidates were vying for this job, would I still be feeling this sudden drain on my motivation? What's really changed?" The truth is that the only factor that has changed is your awareness that at least six others, and probably more, are competing against you. Does this observation make you any less competent, skilled, or experienced than you were when you agreed to the interview? Absolutely not. Your footing to compete at your highest level of ability need not slip an inch. Realizing this, you march into the interview and put everything you've got into getting the position. And, as we're about to see, you'd also do well to believe that you will receive the result soon after you leave the building.

FEEDBACK: THE FASTER THE BETTER

Shifting just slightly to another vantage point, let's talk about the role performance feedback plays in the motivational mix. A strong argument can be made for positive feedback increasing motivation and negative feedback dampening it. But an equally strong argument can be made that negative feedback increases motivation, at least for some of us, because it presents a challenge to overcome. There's little point debating this because, depending on who you are, either argument might apply. What's not nearly as clear is the effect of *when* we receive the feedback—or, more precisely, when we *expect* to receive it.

Let's say that you're preparing for an extremely important test that you and roughly one hundred other classmates will be taking in a week (or, if you prefer, an executive training program exam, or certification exam—pick your poison). A few days before the test, you find out that your instructor will be going on a trip not long after the test is over and will be providing written and verbal feedback to the students within a day of the test. This is unusual, because ordinarily the instructor waits a week or more before providing feedback. About half of the class finds out that they'll be getting rapid feedback, and the other half thinks they won't receive feedback for several days, per usual.

Which group is more likely to perform better on the test?

That question was investigated by University of Alberta researchers Keri Kettle and Gerald Haubl, who hypothesized that the mere anticipation of proximate feedback would result in better performance on a test. Previous research has shown that when feedback is rapid, the threat of disappointment increases.[3] The desire to avoid the dreary feeling you get when you fall short of expectations is a potent motivator to perform well. Students were recruited into the study by emails sent one, eight, or fifteen days prior to a nerve-wracking test of their performance: making a public presentation. The students were reminded of their presentation date and also told when they would receive a grade, which would be provided as a percentile score (e.g., 90th percentile, 70th percentile, etc.). Then they were asked to predict their performance by selecting a grade rank from out of ten possible percentile grades. In all, 271 people ranging in age from eighteen to thirty-two participated in the study.

You Can Be Afraid to Lose, Just Don't Lose Perspective

Sweaty palms and upper lips, fidgety fingers and bouncing knees, frantic, racing thoughts—all are signs of emotional tumult when facing the risk of loss, and all seem involuntary. But a study indicates that we can influence the degree of emotional reaction and our level of loss aversion. The solution, in short: Think like a stock trader. Seasoned traders are careful not to lose perspective when facing potential loss. They view loss as part of the game, but not the end of the game, and they rationally accept that taking a risk entails the possibility of losing. Researchers wanted to investigate whether cognitive-regulation strategies (i.e., strategies to change thinking, like those used by traders) could be used to affect loss aversion and the physiological correlates of facing loss. Subjects were given $30 and offered a choice to either gamble the money, and potentially lose it, or keep it. They could theoretically win up to $572 or lose the $30 and be left with nothing. The outcomes of their choices were revealed immediately after the choice was made (e.g., "you won"). Subjects completed two full sets of choices (140 choices per set). During the first set, subjects were told that the choice was isolated from any larger context ("as if it was the only one to consider"); during the second set, subjects were told that the choice was part of a greater context ("as if creating a portfolio")—in other words, the introduction of "greater context" (taking a different perspective) functioned as a cognitive-regulation strategy. The researchers conducted this study twice: In the first, they observed behavior; in the second, they observed behavior and administered a skin conductance test (a measure of sympathetic nervous system activity) to measure level of emotional arousal. The results: Using the cognitive-regulation strategy had the strong effect of decreasing loss aversion. Most important, only individuals successful at decreasing their loss aversion by taking a different perspective had a corresponding reduction in physiological arousal response to potential loss. So cognitive regulation led to less loss aversion, which led to less sweat on the upper lip.[4]

The results were consistent with the hypothesis: Participants who anticipated more rapid feedback scored the highest on the test. The surprising part was how significantly different the grades were for each group. Students who thought they'd receive rapid feedback performed 22 percentile ranks higher than students who thought they wouldn't receive feedback for several days, and this held true across the full range of scores. At the same time, predicted performance went in exactly the opposite direction. Students who predicted they would perform the best actually performed the worst; students who predicted they'd perform the worst did the best.

The reason for these results is that the students who feared disappointment the most (those who thought they'd receive immediate feedback) were more powerfully motivated to do well, while simultaneously reducing their personal expectations of performance to prep themselves for bad news. In other words, motivation to perform well and pessimistic expectations are not mutually exclusive. In fact, they seem to get along famously.

The other students saw disappointment as a more distant possibility and were consequently less prepared for the test, even though they thought they'd do just fine. The takeaway here seems to me very practical: When you're about to face a test of performance (in any walk of life), imagine that you'll receive feedback right away and act accordingly. The proximity of potential disappointment will keep you sharp and ready to perform.

And don't feel bad if a bit of pessimism slips in to help you brace for impact. It's best to view that pessimism as your brain's way of alerting you to the possibility of failure—an unpleasant jolt to cerebral happiness. As we've discussed, that's an alarm you can certainly choose to honor, but doing so isn't going to allow you to reach the level of performance you're striving for.

STARK-NAKED COMMITMENT

So far we've talked about the sort of achiever you are, the role of competition, and the effect of feedback timing—and in each case we've seen that motivation tips up or down depending on a slew of variables that we can identify—and in doing so give ourselves a better chance of capturing the magic. Now let's talk about whether going public with our intentions—as many motivational programs advise—really does amp up our motivation. For this we'll use an example taken from the upper echelon of America's national obsessions: losing weight.

Several of the most popular weight-loss programs operate on the public commitment principle. Individuals are challenged to state "publicly" (which may simply mean in front of a small weight-loss group) that they want to lose so much weight in a given time period. The commitment hinges on social pressure working against the possibility of failure. If someone doesn't succeed, or at least make substantial progress toward the goal, everyone will know it.

On the face of it, this principle seems sound, since no one wants to be publicly embarrassed or viewed as a hypocrite. In practice, however, there's a hitch. For the public-commitment principle to operate at full steam, its adherents must genuinely fear the disapproval of others—and that's simply not true of everyone.

A study conducted by researchers Prashanth U. Nyer and Stephanie Dellande investigated how public commitment affects individuals who fear social disapproval—that is, people with high susceptibility to what psychologists call *normative influence* (SNI)—versus individuals who are not as easily influenced by others' opinions (low SNI). It also tested the efficacy of short-term versus long-term public commitment, as well as no public commitment.[5]

Two-hundred and eleven women between the ages twenty and forty-five were recruited for the study. They signed up for a sixteen-week weight-loss program designed to help people lose fifteen to twenty pounds and maintain weight loss over time. All subjects completed questionnaires that gauged SNI level and personal weight-loss motivation. Subjects were then randomly separated into three groups: long-term public commitment, short-term public commitment, and no public commitment. Those in the long-term group wrote their names and weight-loss goals on index cards that were publicly displayed in the fitness center for the full sixteen weeks of the program. Those in the short-term group did the same, but the cards were displayed for only the first three weeks. Those in the no-public-commitment group did not fill out cards.

At the conclusion of the study, the effect of long-term public commitment was evident. Those in the long-term group lost significantly more weight than the short-term and no-commitment groups. At the sixteen-week mark, subjects in the long-term group had, on average, exceeded their goals to the tune of 102 percent, while the short-term group achieved an average of 96 percent success and the no-commitment group reached only 88 percent.

The effect of SNI level was also evident. Subjects in the long-term group that tested as having low SNI—in other words, low susceptibility to social pressure—achieved an average of 90 percent of their weight-loss goals. In contrast, individuals who tested as having high SNI exceeded their weight-loss goals by a significant margin: an average of nearly 105 percent.

What this study tells us is that in general the public-commitment principle produces results, especially if the commitment is long-term. But, in the mix of people who make the commitment, those who genuinely fear social disapproval—not a personality trait usually given very high marks—will likely succeed the most. Those who couldn't care less what others think are, ironically, more likely to come up short.

TALKING TO YOUR INNER BOB

Are you the sort of person who routinely tells yourself that you probably can't achieve whatever it is you'd like to achieve? Does the voice in your head—the voice of a brain that craves stability—say things like, "Be realistic, you can't *really* do this." And perhaps, fed up with positive self-talk mumbo jumbo in the media, you think that the only self-talk worth listening to is the "realistic" kind—the kind that tells you how it is.

Whatever your feelings about positive psychology and its many spin-offs, I'm here to tell you that credible research has something to say about all of this—and your little voice should be listening. Research by University of Illinois professor Dolores Albarracin and her team has convincingly shown that those who *ask* themselves whether they will perform a task generally do better than those who *tell* themselves that they will.[6]

But first, a slight digression. If you have young kids or even early teens, you may be familiar with the TV show *Bob the Builder*. Bob is a positive little man with serious intentions about building and fixing things. Prior to taking on any given task, he loudly asks himself and his team, "Can we fix it?" To which his team responds, "Yes we can!" Now, compare this approach with that of the Little Engine That Could, who's oft-repeated success phrase was, "I think I can, I think I can . . ." In a nutshell, the research we're about to discuss wanted to know which approach works best.

Researchers tested these two different motivational approaches with fifty study participants, first asking them to either spend a minute wondering whether they would complete a task or telling themselves they would. The participants showed more success on an anagram task (rearranging words to create different words) when they *asked* themselves whether they would complete it than when they *told* themselves they would.

In another experiment, students were asked to write two seemingly unrelated sentences, starting with either "I will" or "Will I," and then work on the same anagram task. Participants did better when they wrote "Will" followed by "I," even though they had no idea that the word writing related to the anagram task. In other words, by asking themselves a question, people were more likely to build their own motivation than if they simply told themselves they'd get it done.

The takeaway for us is that little voice has a point—sort of. Telling ourselves that we can achieve a goal may not get us too far. Asking ourselves, on the other hand, may bear significant fruit. Retool your self-talk to focus on the questions instead of presupposing answers, and allow your mind to build motivation around the question. Or take a shortcut and just remember the anthem of Bob the Builder.

Parallel to the motivation issue is commitment to goals, and the restraint required to reach them. We'll visit those topics in the next chapter.

Chapter 7

WRITING PROMISES ON AN ETCH-A-SKETCH

"I can resist everything except temptation."
—OSCAR WILDE, *LADY WINDERMERE'S FAN*

THIS TIME I *REALLY* PROMISE

Robert is a well-regarded pharmacologist who runs a clinic for diabetes patients in Orlando, Florida. Many patients of this clinic do not want to begin a regimen of insulin shots (due to the obvious reasons: pain, and the stigma of having to take them) even though they have failed to control their blood glucose by other means. Weight loss offers the only real hope of eliminating their need for insulin, and patients will often beg for another chance to lose weight so they can avoid the need for shots. Nearly 100 percent of the time, in the pharmacologist's experience, this cycle continues indefinitely (or until he stops it): At each appointment the patient will beg for another opportunity to begin losing weight and will express his or her seriousness "this time."

But very few ever follow through. There is always a reason why they weren't able to do it the *last* time, but they should be able to eat better/exercise more now that some small circumstance has changed ("we had family in town," "I had a stressful situation with my sister," "my job changed," "I had a cold," etc.). The cycle continues, and their health worsens.

Robert has an interesting technique he uses with patients caught in this cycle. He "stops the game," so to speak, by taking the ball from the patient and initiating a negotiation. First, he clarifies with them whether they truly understand that they only have two options: lose weight or take insulin shots. When this is confirmed, he then asks them to commit to an amount of weight they intend to lose in the next month. The patient must come up with the number. He then asks them to commit to a particular day upon which they will be weighed to prove they lost the weight. Again, it's up to them to pick the day. The important part about this step is that the patient is empowered to establish the goals. Then he asks them to firmly commit, then and there, that if they do not reach the goal by the date, they will begin taking insulin shots. Almost everyone at this point agrees to accept the terms of the negotiation.

What Robert knows from years of experience is that 90 percent of the patients will still not lose the weight, but because they (and not he) established the goals and parameters, they will more easily accept the consequence. In effect, he is making them face their "game," and by doing so stops the cycle. If a patient does lose the weight, then all the better—but if not, at least he or she will willingly receive the necessary treatment.

This is one example of chronic self-restraint failure, and also of an intervention strategy to prevent the consequences of the failure from causing even worse outcomes. If Robert took a hands-off approach to his patients' situations, the cycle would no doubt continue, and for many of his patients the results would be catastrophic.

Most of us don't have the luxury of relying on a trained clinician like Robert to stymie the results of our self-control breakdowns, and what's worse is that our brains are not natively tuned to offer much assistance. As any yo-yo dieter knows, when you severely cut back on calories, your brain responds by decreasing your caloric burn to keep you from starving to death (the fact that you were never in danger of starving to death doesn't matter; our brains react with tried-and-true solutions spanning centuries). When you deprive your body of carbohydrates, as with a restrictive protein diet, you will eventually crave them to the point of hyperloading

WRITING PROMISES ON AN ETCH-A-SKETCH

and gaining more weight than you lost on the diet. For every aggressive move we make, our brains have a countermove. Ironic, isn't it, since *all* of the moves originate in the same place? Let's explore a few restraint quirks and follies and see if we can't get a better handle on the insanity.

ALL THE RESTRAINT YOU CAN EAT

For six months you have worked really hard to stick to a diet, and it's paying off. Not only have you lost weight, but now more than ever you're better able to restrain your impulse to eat fattening foods. Your friends are telling you how impressed they are with your resolve, and truth be told you are feeling pretty damn good about yourself as well.

Which is why, around month seven, you decide that your impulse control is sufficiently strengthened that avoiding being around ice cream, nachos, chicken wings, soda—and all of the other things you used to eat out with your friends—is no longer necessary. You've spent half a year changing the way you think about food, and it worked. Maintenance won't be difficult with a new mind-set. Time to live again!

I probably don't have to end this story for you to know how it turns out. It's a classic tragedy with which many of us are already too familiar. Pride comes before a fall, but even more often it's our sense of inflated self-restraint that precedes a tumble into relapse.

Researchers from Stanford University, Northwestern University, and the University of Amsterdam teamed up to investigate the dynamics underlying why we repeatedly convince ourselves that we've overcome impulsiveness and can stop avoiding our worst temptations.[1] This particular tendency toward self-deception is what psychologists call *restraint bias*, and four experiments were conducted under this study to test the hypothesis that it's rampant in our bias-prone species.

In one of the experiments, people walking in and out of a cafeteria were approached with seven snacks of varying fattiness and asked to rank the snacks from least to most favorite. Once they finished ranking, participants were told to pick one snack, and further told that they could eat it at any time they liked, but if they returned the snack to the same location in one week they'd receive five dollars and could also keep the snack. After choosing the snack, participants indicated if they would return it for the money, and then filled out a questionnaire that assessed their hunger level and impulse-control beliefs.

Participants who were walking into the cafeteria said they were hungry, and those leaving said they were full; so the first evaluation was whether those leaving with full stomachs would indicate stronger impulse-control beliefs—and they did. The next evaluation was whether the not-hungry participants claiming the most impulse-control would choose the most tempting (and most fatty) snacks. They did. Finally, would those who selected the most tempting snacks be least likely to return them a week later? Indeed, they were.

In another experiment, heavy smokers were asked to take a test to assess their level of impulse control. The test was bogus, designed only to label roughly half of the participants as having a high capacity for self-control and half as having a low capacity. Being told which label they earned seeded participants with a self-perception in either direction.

Participants were then asked to play a game that pitted the temptation to smoke against an opportunity to win money. The goal of the game was to watch a film called *Coffee and Cigarettes* (filled with scenes of people smoking) without having a cigarette. They could select among four levels of temptation, each with a corresponding dollar value: (1) keep a cigarette in another room: $5; keep a cigarette on a nearby desk: $10; (3) hold an unlit cigarette in their hand throughout the film: $15; (4) or hold an unlit cigarette in their mouth throughout the film: $20. Participants earned the money only if they avoided smoking the cigarette for the entire movie.

As predicted, smokers who were told they had high self-control exposed themselves to significantly more temptation than those told they had low self-control. On average, low-self-control participants opted to watch the movie with a cigarette on the table; high self-controllers opted to watch with a cig in their hand.

The result: The failure rate for those told they had high self-control was massively higher than for the low-self-control group, to the tune of 33 percent versus 11 percent. Those who thought themselves most able to resist temptation had to light up three times as much as those who suspected they'd fail.

Can You Control Yourself Better Than a Chimp?

Coping with impulsivity is a more complicated ability than it appears, and for a long time we thought only humans could do it. If you put candy in front of a group of children and tell them that if they can resist eating it they will eventually receive even more candy, a few interesting behaviors will follow. Some of the kids will try to distract themselves by playing with toys or drawing as a way to cope with the frustration of delaying gratification (some of the kids will just give up and grab the candy). This is quite advanced problem-solving behavior, which is why it was surprising to find out that chimpanzees can do it, too. Researchers showed a group of chimps a small pile of candy that was accumulating more candy as time passed by, but the candy was inaccessible to them. Then they gave the chimps a set of toys. Every so often they would allow the chimps to have access to the candy. Several of the chimps caught on that the longer they waited, the more candy would accumulate, so they distracted themselves with the toys to avoid grabbing the candy until they could get a hefty amount. The chimps became intensely focused on the toys when the candy became accessible, showing that they really were diverting their attention so they could get a bigger reward later.[2]

One way to view these results is as reinforcement of a very old cliché: We're our own worst enemies. Restraint bias has a place high on the list of biases we stumble on routinely, and tripping on it once is no guarantee of not doing so again, and again . . . and maybe again. Dieters relapse, smokers relapse, anyone with anything approaching a compulsion relapses—usually more than once. This study suggests that part of this repetition is due to thinking we can handle more than we can.

Another takeaway is that an entire industry is based on bolstering impulse control. Self-help books and motivational speakers aplenty play on a dubious concept that there's a gold ring of restraint we all can reach—just follow X system to get there. But what this study suggests is that even if you think you've arrived "there," you'll eventually find out that "there" never existed. You were sold a mirage in the form of an inflated self-perception of restraint. Sorry, no refunds.

OUTSOURCING SELF-CONTROL

Common among couples who have been together for several years is the phenomenon of being able to finish each other's sentences. After a while it becomes one of those "funny things couples do." And if you notice, the phenomenon is especially true when one of the partners is trying to remember something; the other partner fills in blanks in the discussion with pieces of memories the other can't recall. Psychologists studying how this works call it *transactive memory*—which simply means that over time relationship partners are able to rely on each other to remember things.[3] They become a memory duo, sharing certain memories that neither one of them can reconstruct in total. In light of the brain's energy conservation strategy (discussed more in chapter 10), this arrangement makes a lot of sense.

As it turns out, something similar happens with self-control—and the news about this is both good and bad. The good news is that "transactive self-control" evidences a strong bond between partners and probably contributes to reaching long-term goals—such as the discipline to get a degree. The bad news is that it appears to undermine short-term self-control goals—such as losing weight. That was the conclusion from a study conducted by psychologists at Duke and Northwestern Universities, testing the pros and cons of outsourcing self-control to romantically tied partners.[4] The study showed that when someone expects a level of support from a partner to stick to a diet, for instance, that person will actually decrease their energy expenditure. The effect was especially strong for participants who were already worn out from other energy drains. Alongside that result, participants who relied on a partner for help with their studies procrastinated more than those who went it alone. The reason, returning to a common theme, is energy conservation.

Our brains are huge energy consumers (15–20 percent of our daily caloric intake fuels the brain) but stingy energy users. If an external resource is available to draw on instead of using stored energy, you can be certain the brain will want to tap it. Ironically, as this study suggests, doing so in the short term damages self-control by weakening our resolve to try as hard as we otherwise might. (This result should not, by the way, be misconstrued to contradict the public commitment findings discussed in the last chapter. In that case, the commitment is reinforced by others *holding us accountable* to achieve a goal, not helping us in any substantial way to get the work done.)

FLAVORS OF IMAGINATION THAT DENY TEMPTATION

Think about chocolate. A nice big bar of especially delicious Belgian chocolate on a table in front of you, and it's all yours. Imagine unwrapping it, smelling the sweet chocolate aroma, breaking off a piece, and bringing it to your mouth for a taste of ecstasy.

For most of us (aside from those who inexplicably don't like chocolate), that description will kick up a craving for chocolate or something sweet. In fact, studies on self-restraint have found that the description need not be nearly so detailed. In one study, simply imagining placing thirty M&M's into a bowl significantly increased how many of the candies participants devoured afterward.[5]

But here's the twist: When participants in the study were told to imagine *eating* M&M's, they actually ate less of them—1.6 times less than the group that imagined placing them into a bowl. The reason seems to be that the brain's response to a conjured image of placing or eating M&M's is much like its response to the real thing. The anticipation of eating the candy drives up temptation, but the image of already chewing and swallowing M&M's drains the energy out of the temptation.

The problem, of course, is staving off temptation long enough to visualize eating the target of our sweet lust before pouncing on it. Far easier said than done.

OH, WHAT THE HELL

Here is something I frequently see, and admittedly do, on business trips. A large group sits around the table at a nice restaurant and a couple of people order several appetizers for the table to share and a couple bottles of wine to start things off in style. The appetizers come along with baskets of bread, also passed around and devoured. Then orders for the meal are taken and everyone gets a salad or soup to begin, followed by a steak or other rich dish. More wine follows. Afterward, most also order dessert and coffee to cap things off, if not a glass of port or grappa.

Before these dinners, I typically tell myself that I will have no more than one small appetizer and pass on the bread. Then I will order a salad with a not-too-terrible dressing followed by a semihealthy dish (at least compared to most) like fish. No dessert. No more than one glass of wine. That's what I tell myself, but when the event begins, something peculiar happens. The appetizers arrive and they look delicious, and I am hungrier than I expected, so I have two or three samplings. Then the bread comes around and I think, "Well, I already ate more appetizers than I should have, so what the hell, might as well have a roll." When ordering the entrée, the "what the hell" effect elevates even more, and I just go ahead and order that juicy steak instead of fish. By this time, "what the hell" is the overriding sentiment and, having blown every self-control commitment so far, I feel fine about ordering dessert with coffee.

Notice the degeneration of control and how with each slipup, the following slipups became easier to make. And it's also important to note that the "what the hell" effect isn't just about self-control, but also about failure to reach goals. Janet Polivy and her research team plumbed the depths of "what the hell" with an experiment featuring two delicious treats: pizza and cookies.[6] The researchers invited 106 female participants to the study, some whom were dieting and others who weren't, under the pretense that they would be tasting and rating a variety of cookies. They were all told not to eat beforehand and were served one slice of pizza when they arrived (exact same size piece for everyone), then asked to sample and rate some cookies.

Here's the twist: A portion of the participants were made to believe that they had received larger or smaller slices of pizza than others. Some of the women got to see another person's slice just before it was carried into that person's separate test room. The slice in question was either one-third larger or one-third smaller than the actual one that the experimental subject was given to eat. In other words, some people were made to think they'd eaten more than the others; although in reality they'd all eaten exactly the same amount.

Then three enormous platters were brought out with piles of oatmeal-raisin, chocolate-chip, and double-chocolate-chip cookies. The participants were told that they could eat as many as they needed to rate the quality of the cookies. What they did not know was that the platters were weighed before and after they were brought out, so the researchers knew exactly how many cookies were eaten. When the cookies were weighed, it turned out that participants who were on a diet and thought they had already blown their calorie-restriction goal ate more of the cookies than those who weren't on a diet—over 50 percent more. On the other hand, when dieters thought they were safely within their calorie limit, they ate the same amount of cookies as those who weren't on a diet.

Again, it's a matter of goals and our perception of how close or how far away from reaching them we are. The farther away we perceive ourselves to be slipping from the goal, the more "what the hell" thinking seeps in—and this tendency applies to a variety of goals besides dieting. Perhaps your goal is to stop smoking. You go two weeks without a cigarette, and one night find yourself at a party with friends, several of whom are lighting up. You think to yourself, *I've made it two weeks, which is pretty good, so I can afford to have one smoke in a social setting.* Later that night, your friends continue smoking and the party is still going strong, and you think, *Well, I've already had one tonight, so what the hell, might as well have another.* Before long you are wrestling with the fact that you have slipped well away from your goal and must start over again.

A DIFFERENT KIND OF CONTROL

So far, everything we have discussed in this chapter has to do with applied restraint (or lack thereof), but there is an entirely different kind of self-control that we engage in without realizing it. First, a little story. Elton is a philanthropist who specializes in raising funds to support children's hospitals all over the world. In some cases, he works with local agencies to develop business plans to build a new hospital, often in a third world country where it is desperately needed. In other cases he works directly with potential funders to help them identify hospitals in need of support and work through the logistics of making sure the money is used effectively. Over the last fifteen or so years, he has helped raise nearly one billion dollars for children's hospitals in fifty countries—the equivalent of helping more than one hundred thousand children get the vital care they needed and would not have received without a fully functioning children's hospital in their region.

Stress Makes You Want It More But Enjoy It Less

Life is lived in loops. Here's one you may know: we experience stress; to relieve the stress we do something pleasurable; when that pleasure exhausts itself, we experience more stress. Sound familiar? Psychologists tell us that when we run in this loop long enough, we may encounter something called *anhedonia*—the inability to experience pleasure from things we'd normally enjoy. Does binging on chocolate do it for you? Binge long enough and it probably won't. To this stinging realization we can add another, and— apologies ahead of time—it's also a bit of a pill. Researchers have shown that not only does stress predispose us to wanting pleasure, it makes our desire for it drastically out of proportion to our enjoyment. The reward never reaches the level of our *want*.

To demonstrate this, researchers recruited two groups of study participants—all of whom were chocolate lovers—for some fun with water and sweets. Members of the first group were made to experience stress by holding their hands in ice water (a well-tested means of inducing stress in psych research) while they were observed by the researchers. Those in the other group placed their hands in lukewarm water. After a little while, both groups were told to squeeze a handgrip that, they were instructed, would give them a nice stout whiff of chocolate.

As you might predict, the stressed group squeezed considerably harder for their chocolate—*three times* harder. Having received their dose of reward, the groups were then asked to rate their satisfaction. You may think that the group desiring the chocolate with three times the intensity would rate it proportionally higher, but in the end the groups' ratings were really no different. Using stress to spike desire did nothing to increase enjoyment. Stress seems to flip a switch in our brain that makes us want the pleasure more—driving us to expend more effort to get it—even though the effort won't deliver the pleasure premium we're seeking.[7]

Ethical questions crop up during the course of Elton's work. Certain investors aren't as interested in supporting children's health as they are in courting favor with local authorities, who, in light of the investments being made, are often more willing to overlook less commendable things the investors are involved in. Sometimes this means waving regulations, reducing taxes and fees, or, in more extreme cases, ignoring blatantly illegal activity. The philanthropist is usually aware that these things are happening, and sometimes Elton's fee gets a boost if the activity in question is especially sordid. He knows, of course, why he's getting more money, and he could turn it down. In fact, he could stop the deal altogether or at least refuse to participate. He doesn't. In his mind, the scale is balanced. If he didn't participate in the deals, the children's hospitals wouldn't be built and maintained. If he ignores unethical behavior, and even occasionally takes what amounts to a bribe to keep quiet, that's okay, because on the other side of the seesaw he is doing noble work.

Elton's thinking and behavior illustrates what psychologists in recent research have dubbed the "moral self-regulation effect"—the tendency to conduct a balancing act in our lives by doing something moral in one case to offset doing something wrong (or doing nothing at all) in another.[8] When we do a moral act to offset an immoral one, we are engaged in "moral cleansing." When we do nothing, or perhaps something perceived as immoral (because we feel like we have enough in the moral bank account to get away with it) we are engaged in "moral licensing."

A great deal of "green" marketing is predicated on the assumption that people will buy a green product to make themselves feel better about moral deficiencies in other parts of their lives. Other forms of green messaging, such as expensive hotels asking patrons to reuse bath towels, use this same dynamic; nothing is offered to the patron in return for not using the towels other than a feeling of "doing good for the environment"—a feeling that will offset a sense of moral deficiency for not recycling at home (as an example). The hotel, meanwhile, benefits from reduced costs, which add directly to its bottom line. (My purpose here is not to disparage worthwhile environmental campaigns, but only to show how the moral self-regulation effect is applied in real situations.)

The subtlety of this effect is on the level of background noise—it's happening all of the time and we rarely give it a second thought. The important point to remember is that we use mechanisms like moral self-regulation to gain balance and feel more at ease with our place in the world. Balance makes the brain happy, and feeling at ease is whipped cream on the sundae.

We will come back to self-control machinations later in the book, but now we will turn to one of the more misunderstood cycles of thinking any of us experiences: the cycle of regret, and all emotions pertaining to it.

Chapter 8

WANT, GET, REGRET, REPEAT

*"I see it all perfectly; there are two possible situations—
one can either do this or that. My honest opinion and my
friendly advice is this: Do it or do not do it—you will
regret both."*

—SØREN KIERKEGAARD,
*BALANCE BETWEEN ESTHETIC
AND ETHICAL*, VOL. 2: "EITHER/OR"

GOING Z INSTEAD OF Y . . . OH WHY?

Madison was never entirely sure that she wanted to become
a lawyer. After spending two years at a law firm, she was
becoming absolutely certain that she should *never* have become one.
Her expectations for the profession may have been distorted, or perhaps
she simply had a naïve view of the law from the start. She was willing
to concede these points, but doing so was not making her day-to-day
existence any easier. To continue on in a profession that failed to inspire
any degree of passion and commitment was a horrible prospect, but
how could she possibly change direction after years on this road, to say
nothing of the huge financial investment she had made to get there—one
that would take years to pay off? For Madison, every day was a draining
struggle with regret.

After digesting a vignette like that, I'm sure everyone reading recalls
with discomfort similar situations in their lives—perhaps not this

extreme, but no one living who makes decisions escapes the pain of regret in at least one area of life. Most certainly more than one. Our brains experience regret as a form of loss, and as we've seen, avoiding loss makes our brains happy. The problem is that avoiding regret is rarely possible, and attempting to do so is perilous business in its own right. We also fail to realize that regret is not one thing; it manifests in overt and covert forms that materialize in our brains as varying levels of loss. For all of the fear and loathing it generates—to say nothing of the thousands of songs and poems it inspires—regret is a deceptively complicated topic; we will attempt to unravel at least part of the enigma in this chapter.

WILE E. COYOTE, FAUX GIRLFRIENDS, AND EBAY

For those of us who grew up watching Looney Tunes cartoons from Warner Brothers, the Road Runner was a Saturday-morning staple. The premise was simple: An especially lucky road runner (a fast desert bird) is relentlessly pursued by a desperate coyote. No matter what the coyote does, including using every contraption imaginable from the Acme Corporation, the road runner manages to escape him. The first of these cartoons aired in 1949, and since then the inexhaustible duo have appeared in countless episodes, ranging from shorts to full-length features. With such a simple premise and just two characters, it's interesting to wonder why the cartoon has been so popular for more than sixty years. We'll come back to that question in a moment.

Director Steven Soderbergh is known for movies that take on gritty topics without pulling too many punches. His style is to give the audience an "on the scene" perspective by using handheld cameras, moody lighting, and by putting viewers in the middle of the action. In 2009, he turned his attention to high-end prostitution in a film called *The Girlfriend Experience*. The film follows a successful call girl as she tries to take her career to the next level. What she offers, as the title suggests, is the experience of having a beautiful girlfriend without having to manage an actual relationship. Her customers are wealthy, often married, and each seeks a different experience from her. Some want sex; others want to talk, with various shades between. At the same time, she is attempting to manage an ongoing relationship with her oddly understanding boyfriend, though the relationship steadily falls apart as the movie goes on. At one point, the main character thinks that she may want to have a true relationship with one of her customers, and he agrees that they should try and see where it goes. Predictably, it goes nowhere. As soon as the chasm between the "girlfriend experience" and an actual relationship is breached, the fantasy is over.

You would have to search long and far to find someone who has not bid for something on eBay. Volumes of research have been written about the online auction phenomenon. Of particular interest is what drives people to continue to bid on items even when the price eventually exceeds the value of the item and/or what the bidder was originally willing to pay.[1] Many factors have been cited—and no doubt the reasons are not the same for all people—but the one consistently mentioned factor is that the allure of anticipating the win is immensely powerful. It is so powerful, in fact, that after focusing for days on winning an item and fighting off those who would nab the prize, bidders often feel an emotional letdown. They won the item, and in a few days it will arrive in the mail—excitement not included.

As I am sure you have noticed, the three examples I just described share a central theme: the power of wanting trumps the satisfaction of getting. For a decades-old cartoon featuring nothing more than a coyote and a bird to remain popular, the tension of wanting has to be preserved. It doesn't "work" for the coyote to ever get the road runner. People love the cartoon because Wile E. Coyote wants the prize with every fiber of his being and forever fails to reach it. For the girlfriend experience to remain a profitable endeavor, it cannot become the "girlfriend for real experience." The energy (and money) is derived from the tension of wanting something that will never materialize. And for many bidders on eBay, it is the thrill of the hunt that motivates higher and higher bids, even when the same item could be bought elsewhere for less money and with less time required to get it. Obtaining the thing in question often leads to a dull, hopeless feeling peppered with confusion. That is the *regret of getting*—a feeling of loss that, if it could speak, would ask, "Now what?"

SINGING THE HABITUATION BLUES

And as you might expect, our brains have an answer to that question: Target a new reward. That answer makes sense because the brain's reward system is structured to drive us to continually seek beneficial rewards, be they food, water, sex, shelter, or proxies for these, like money and all that it can provide. The problem, of course, is that when we get the thing we wanted, the game is over. On top of this, something psychologists call *habituation* begins to set in and we get used to the thing in question in a mere matter of days or weeks.[2] An example of this process that we see every day is new technology purchases; whether it's a new computer, video-game console, or smartphone. First, we feel elation in anticipation of buying the item. When we finally do buy it, that high-pitched elation gives way to a more grounded "liking" of the item, which, over the next few weeks, dissipates into a neutral appreciation. That stage may last for a while, but eventually the item becomes just another possession in your collection with certain usefulness (if it's even still useful). The initial feeling of elation can never be totally recaptured for that item once the cycle comes full circle. What's the brain's answer? Focus on a new reward and get back to elation. (For more on the brain's reward system, take a look at Special Section 2 at the end of the book.)

In fact, regret can set in even before we finalize getting or doing that which we've anticipated. Research shows that regret is one of the most influential factors in decision making because we feel it so strongly once the anticipation of reward starts settling into something much less exhilarating. The drop-off often begins even before the decision is finalized.[3] Several examples illustrate the process. For instance, consider a man or woman who is divorced and remarried multiple times. Psychology research on habituation in relationships suggests that many people simply never overcome the elation drop-off once the reality of their commitment sets in. Regret fills the void, and the marriage suffers until it ends. Not by coincidence is the divorce rate for second marriages higher than it is for first marriages; attempting to recapture the anticipation of reward often sets off a new cycle of habituation and regret.

Another example that most of us can relate to is purchasing a new car. It's not uncommon for regret to set in even before the deal is finalized, because once the would-be buyer enters the showroom, he or she is faced with comparisons between the car they intend to buy and all of the other cars on display. With each observation of features other cars have that the intended car lacks, the door of regret is cracked open a bit more. When the car is purchased, postcommitment comparisons begin with other cars on the road, in magazines, on TV, and so forth. In this case, habituation and regret move along parallel tracks.

Or consider moving to a new city and starting a new job. It's common to believe that the new place or position will be significantly better than what we have become accustomed to. "Newness" has a sort of mystical draw, even though it's incredibly short-lived and rarely meets expectations (because those expectations are colored by an unattainable desire for ongoing elation). When we move to the new place and start the new position, the same cycle of regret sets in as anticipation of reward drifts off—hence the quirky but insightful axiom, "Wherever you go, there you are."

Regret Death Match: Wanting versus Liking

Let's say that you're on eBay and you see an auction for something you *really* want. You end up having to fight for it right down to the wire, but eventually lose to a last-second sniper. Annoyed, you go prowling around and find the same item for more money as a "Buy it Now." Without hesitation, you buy it, paying a substantial premium over the ending price from the auction you just lost. A week later, the item arrives at your house. You open the box and are elated, right? Wrong. In fact, you can't even recall why you liked the thing so much to begin with. That night, you put it back up for auction on eBay.

Does this make any sense to you?

A study published in the journal *Psychological Science* suggests that this scenario, with whichever elements you'd rather sub in, isn't only plausible, it's predictable, and it has much to do with the peculiar love–hate relationship between wanting and liking and the regret fallout it produces.

As we've all experienced, when you really want something but are prevented from getting it, you want it all the more. This is even truer of relationships than objects. The jilted lover syndrome is a Hollywood mainstay because just about everyone can relate. Researchers started with that well-known phenomenon and wanted to know how they could create a "counterdrive" dynamic between wanting and liking—that is, causing someone to pursue her "want" even after her "like" is gone. In one experiment, participants were offered an opportunity to win a prize they said they wanted. When they failed to win (in other words, when their "want" was jilted), they were offered the opportunity to buy the same item for more money than it cost those who won it. By a significant margin, those who were jilted did exactly that. But when then asked if they'd like to trade the item away, most of the jilted crew said "yep, take it." In another experiment, participants were given the opportunity to win Guess brand sunglasses. Those who were jilted and didn't win the sunglasses were then presented with an opportunity to choose between a Guess wristwatch and a Calvin Klein wristwatch. Most of those who were jilted chose the Guess wristwatch. You might think that's because they really like Guess products, right? Nope. When asked for their evaluation of Guess wristwatches, they rated them surprisingly low. What's going on here? The research team believes that when our desire is stoked, we're in an intense emotional go-mode. But when we're jilted, the intensity of our emotion goes negative, and that negativity rubs off on the object of our original desire, catalyzing feelings of regret. The weird thing is that the same intensity still pushes us forward to get (or try to get) the thing we wanted even when we are beyond liking it.[4]

THE COUNTERFACTUAL CONUNDRUM

For all its negatives, regret actually serves an important adaptive function. Without it, our ability to learn, change, and improve would fall short of what our species has needed to survive and thrive. Regret as a learning tool happens through something called *counterfactual thinking*—a dynamic with two razor-sharp edges.[5] When we look back on a decision and think, *If I had done A instead of B, then I wouldn't have to deal with horrible C,* we are engaging in counterfactual thinking. We are imagining decision components "counter" to those that actually transpired and arranging them in such a way that they result in a different outcome. Madison, from this chapter's introduction, thinks to herself that if she had pursued her original goal of becoming a graphic designer (a choice counter to what she really pursued) and not been so frightened by the possibility of not finding a good job after graduation, then she wouldn't dread every day that she must now function as an attorney.

Counterfactual thinking involves a hefty dose of dwelling on what we think *should* have happened, and that can be immensely frustrating. Every situation we'll ever face has alternative realities that might have transpired had we chosen differently, or had the elements of the decision been presented to us differently. Some of those might have been under our control; some surely were not. From a learning standpoint, counterfactual thinking can help because the next time we face a similar situation we will have the results of our counterfactual thinking in mind and will not make the same errors again (at least in theory). But from an emotional-health standpoint, spending too much time in the counterfactual examination room can lead to serious consequences.[6] If we allow ourselves to dwell on a bad decision and everything else we could have done to avoid a bad outcome, negative emotions will overshadow the learning benefits. For those suffering from depression, obsessive counterfactual thinking is like propane gas feeding a fire.

Unfortunately, beating ourselves up over poor decisions is a hard habit to break. Our brains revisit these decisions because learning from error is central to a happy brain's reason for being. It's an adaptive trait that we can't afford to live without. The problem is that we lack an internal governor to regulate how much of this learning process to indulge, and eventually what could help us ends up hurting us.

How Stores Manipulate Regret

Because regret is such a powerful dynamic, it's one of the favorite tactics used by cajolers of every stripe to get us to see things their way. For example, a salesperson will set up the factors in a buying decision such that it appears making one choice instead of the other will amount to immediate and lasting regret. This is how stores sell product insurance plans, which are really just sources of pure profit because seldom does anyone use them. Maybe you've heard this line when purchasing a new TV or computer: "The insurance plan on this item is less than 5 percent of the total cost, and you'll have peace of mind that your investment is covered if, for example, a power surge blows out the equipment." Power surges are a favorite tool to elicit counterfactual thinking because they're so mysteriously dangerous. You never know when one might strike. "Imagine," the pitch continues, "if you don't buy the plan and a year from now your $2,000 TV is destroyed by a sudden power surge." The customer's mind races to the future, a smoldering TV in his living room, $2,000 gone just like that. A mere 5 percent now to avoid that terrible fate later? Sure, it's the smart thing to do. Think again. Odds are, you are paying for nothing.

AND THE GOOD NEWS?

So what can we do about it? Recent research suggests that we can "train" our brains to temper anticipation of reward without squashing it altogether (certain brain injuries can result in loss of anticipation of reward;

it's not something most of us would want to lose). Assuming a partially detached stance from our own thinking (a metacognitive stance, one that enables *thinking about our thinking*) enables us to examine the elements of our drama without being consumed by the drama. Even short periods of thinking about our thinking can help, and journaling our thoughts as we do so is especially useful. Encouraging a constructive inner dialogue—one that doesn't include self-destructive and self-demeaning talk—can help us examine different ways of thinking through the issue, while maintaining a focus on our (non-counterfactual) present realities.

At the same time, psychology research indicates that when we're making decisions about what we want, focusing on experiences involving friends and family yields greater long-term rewards and far less regret than purchasing material items. The issue is more complicated for relationships, but solid research has shed light on this complex topic as well.

More on those suggestions in chapter 15. For now, we must move on to examine the social sandbox in which we all play, and what research tells us about the influence we possess and the influences that possess us.

Part 4

SOCIAL EBBS AND INFLUENTIAL FLOWS

Chapter 9

SOCIALIZING WITH MONKEYS LIKE US

"Society is a masked ball, where everyone hides his real character, and reveals it by hiding."
—RALPH WALDO EMERSON,
CONDUCT OF LIFE

MONKEY SEE, MONKEY DRAMA

Professor Laurie Santos is one of the leading primatologists in the country. As the director of the Yale University CapLab (aka the Comparative Cognition Laboratory), she has developed an understanding of capuchin monkey social systems that is challenging many of our long-held assumptions about monkey and human distinctions. At times, she explains, watching the capuchins is just like watching a human soap opera (without the cheesy dialogue). Monkeys display jealousy, grief, worry, joy, and a range of other emotions that we used to think were exclusive to humans. They also cheat on their partners, steal, and alienate others, just like humans do. As it turns out, the dynamics of monkey society are not unlike our own—in some ways, they are startlingly similar.

What is not the same, Santos points out, is how we and our capuchin cousins navigate our way through our respective social landscapes. The reason for this disparity is that natural evolution and cultural evolution move at entirely different speeds. Santos comments: "Culture moves much faster than natural selection; too fast for natural selection to ever catch us up biologically."[1]

Another way to state that observation is that our social infrastructure is far more complex than natural selection could prepare us for. When we observe monkeys, we see a species exhibiting emotional undercurrents similar to our own and addressing them with basic skills that fit the need. If an alien race were to observe us, on the other hand, they would see a species wrestling to manage social complexity that is frequently over our heads, threatening to drown us on any given day.

What this means is that our brains are in many ways at odds with our social environments. Happy brains are protective, predictive, and conservative—not the best fit for human societies that place high value on unpredictability, speed, and consumption. And our social technologies have added many more layers of complexity. If capuchin monkeys could use smartphones and were infatuated with social media apps, we might get a glimpse of how we behave with our technologies by watching them (odds are that what we'd see them do wouldn't be much different from what we do). They'd be monkeys just like us.

Nevertheless, this is where we find ourselves, and our social culture is ours to manage no matter how profound the difficulties—after all, *we* are the crafters of our societies. So let's do some digging.

HI THERE, I'M EVALUATING YOU

We will begin with a core element of social dynamics: first impressions. We all intuitively know the importance of first impressions; from an early age, the mantra "you never forget a first impression" is pressed into our psyches—but what's really going on when we first meet someone that has such a significant impact forevermore? Researchers from New York University and Harvard joined forces to identify what neural systems are in play upon first acquaintance. To accomplish this, the research team designed a novel experiment in which they examined the brain activity when participants developed first impressions of fictional individuals.[2]

The participants were given written profiles of twenty individuals, describing different personality traits. The profiles, presented along with pictures of these fictional people, included scenarios indicating both positive (e.g., intelligent) and negative (e.g., lazy) traits in their depictions. After reading the profiles, the participants were asked to evaluate how much they liked or disliked each person. These impressions varied depending on the value assigned to the different positive and negative traits. For instance, if a participant valued intelligence more than aggressiveness, he or she formed a positive impression of a profile conveying intelligence. During this impression formation period, participants' brain activity was observed using functional magnetic resonance imaging (fMRI). Based on the participants' ratings, the researchers were able to determine the difference in brain activity when they encountered information that was most important in forming the first impression. Two areas of the brain showed significant activity during the coding of impression-relevant information: the amygdalae, which previous research has linked to emotional learning about inanimate objects and social evaluations of trust; and the posterior cingulate cortex, which has been linked to economic decision making and valuation of rewards.

Both of these areas of the brain have been linked to how we determine the value of things (or "value processing," if you prefer). While the line from study results to behavioral inference is never perfectly straight, it appears that this study indicates we're all hardcore value processors even before "Hello" comes out of our mouths. The subjective evaluation we make when meeting someone new includes—to put it bluntly—what's in it for us. This interpretation is not nearly as cynical as it may seem. We are wired to evaluate others in large part on a trust basis, and to our brains, trust is linked to rewards. Both earning another's trust and feeling at ease enough to extend our trust are rewards in the parlance of the happy brain. It makes sense that our brains begin this evaluation from the first moment we make someone's acquaintance.

Reasons to Doubt Your THOMAS

According to neuroeconomist Paul Zak, the essential part of running a con is not to convince the pigeon to trust you, but rather to convince him that *you trust him.* The neurochemical system at play in the con is The Human Oxytocin Mediated Attachment System (THOMAS). THOMAS is a powerful brain circuit that releases the neurochemical oxytocin when we are trusted and induces a desire to reciprocate the trust we have been shown, even with strangers. When THOMAS is engaged by someone who displays trust, we become more vulnerable to the devices of the unscrupulous. The prefrontal cortex, home of our deliberative, and hence more vigilant, faculties, takes a back seat while THOMAS flirts with disaster. The flip side of this coin is that if THOMAS was never engaged, we'd never empathize with anyone or be able to build relationships. Zak's research suggests that about 2 percent of those we encounter in trust scenarios are—using the clinical term—jerks. These people are deceptive, don't stay in relationships long, and enjoy taking advantage of others. They are particularly dangerous because they have learned how to simulate trustworthiness, which makes them psychologically similar to sociopaths.[3]

Another not-so-obvious aspect of first impressions is that the impression we are trying to give influences how we evaluate others.

That's the finding of a study that included hundreds of participants who watched a short film and then discussed it with another participant.[4] Half the participants were given an "impression management goal" to appear introverted, extraverted, smart, confident, or happy. After the discussions, participants rated themselves and the person they had spoken with across several personality traits. Those with an impression management goal rated their conversation partner significantly lower on the trait they were trying to show in themselves, but not on other personality traits. This seems to happen because when we focus on embellishing a particular trait in ourselves, we unconsciously increase the standard for that trait in others—and they usually fall short. So just because someone you're trying to impress doesn't seem as outgoing, gregarious, or confident as you are, don't assume that they truly aren't. It could be that how you are trying to come across has changed your perspective—your brain wouldn't have it any other way.

HOW WE PRUNE OUR NETWORKS

First impressions behind us, let's move on to the dynamics of the most common of relationships: friendships and acquaintanceships. With the dawn of the social-networking age, these relationships are drifting into murkier waters all of the time—not because social networking necessarily makes them any less meaningful, but because virtual interaction makes them harder to pin down. We are going to examine one aspect of this ambiguity—the turnover of relationships over time.

For ages, sociologists have debated whether personal preference or social context holds more sway over how we meet people and the nature of our relationships (would, for example, your husband have become your husband if you'd met him in a bar instead of via your best friend?). Sociologist Gerald Mollenhorst took on the challenge of addressing this question by crafting a study that investigated how the context in which we meet people influences our social network.[5] To his surprise, he found that we lose and replace about half of our friends every seven years, and as a result the size of our social network remains the same over time. Mollenhorst conducted a survey of 1,007 people of ages eighteen to sixty-five years. Seven years later, the respondents were contacted once again and 604 people were reinterviewed. They answered questions such as: Who do you talk with regarding important personal issues? Who helps you with projects in your home? Who do you pop by to see? Where did you get to know that person? And where do you meet that person now?

Mollenhorst found that personal network sizes remained stable, but many members of the network were new. Only 30 percent of the original friends and discussion partners had the same position in a subject's network seven years later, and only 48 percent of all of the contacts were still part of the social network. Mollenhorst also found that social networks were not formed based on personal choices alone. Our choice of friends is limited by opportunities to meet, and people often choose friends from a context in which they have previously chosen a friend. If the pond had fish the first time, why not cast back in? Also, in contrast to research that suggests people typically separate things like work, social clubs, and friends, this study shows that these categories often overlap.

A HAPPY BRAIN'S SOCIAL PREFERENCES

So we see that our social networks are anything but static. Not only should we expect movement in and out of our social circles, but we should also acknowledge the limiting factors that influence them. Another of these factors is the degree to which we feel an affinity with someone not yet in our social circle—the preexisting "inness" or "outness" feeling each of us gets when making a new acquaintance. It comes as no surprise that people tend to prefer others of the same in-group. If, for example, you're a diehard supporter of a political candidate and someone drives by with a bumper sticker endorsing the candidate, you feel a hint of "inness" with that person. If someone drives by with a bumper sticker of the candidate's opponent, you feel a twinge of "otherness" about that person. If asked why, you might say that the first person probably shares many of your views and you're on the same team, more or less. The second driver is showing with the opponent's bumper sticker that she's on the other team. In effect, you feel a sense of in-group trust with the first person that you don't feel with the second.

But why, exactly, trust a stranger any more than another stranger if you don't really know either of them? That question was addressed in a study conducted by researchers from Australian National University and Hokkaido University in Japan.[6] The study began by establishing two possible rationales for group-based trust. The first is stereotyping: People tend to judge in-group members as nicer, more helpful, generous, trustworthy, and fair. The second is expectation: People tend to expect better treatment from in-group members because they are thought to value, and want to further, other in-group members' interests. Study participants were offered a choice between an unknown sum of money from an in-group member or an out-group member (and were told that the in-group and out-group members controlled the amount of money to allocate as they desired). The initial result was that participants overwhelmingly chose the in-group-member option. And, surprisingly, this result held true even when the stereotype of the in-group was more negative than that of the out-group. Good, bad, or indifferent, the stereotype was ignored in favor of group identity. But when participants were told that the in-group money giver didn't know they were part of the same group, the situation changed. When this was the case, participants resorted to making their choice on the basis of stereotype. If the in-group was portrayed negatively, then the participants were more likely to choose the out-group-member option, and vice versa.

This study suggests that when members of the in-group are mutually aware of their "inness," there's an expectation of better treatment than would be received from an out-grouper. But when that awareness is muddied, reliance on stereotypes kicks in. What does this finding tell us about our biases when selecting new members of our social circles? First, it tells us that we often use shaky criteria for making judgments about people. We determine that one stranger is more deserving of our trust than another either because we put unjustifiably high value on their like-mindedness, or we simply default to a stereotype. Neither of these tendencies speak especially well of themselves, though they evidence basic tendencies that are in all likelihood neurally imprinted. In fact, neuroscience research has identified neural structures in our brains correlating to our social biases—so there is worthy evidence that social bias is, at least to some degree, wired into our noggins.[7]

I Can Hear It in Your Voice

Business management and sales training often include sessions on *how* to say what you want to say—the tone-of-voice packaging of the message. This is partly intuitive; the pitch and weight of one's tone influences the listener—soft to cajole, loud to command, so forth. But regardless of tone, clearly some people's voices carry more influence than others. We seem to be equipped with a way to detect the level of confidence embedded in others' voices, and even a loud tone—if lacking the confidence intangible—isn't likely to cause much more than irritation. Research from McGill University put a finer point on the confidence issue by tracking changes in listeners' brain activity when hearing someone make a statement. The results suggest that the brain detects and assesses confidence in another's voice in as little as 0.2 seconds. Researchers outfitted the heads of a group of volunteers with 64 electrodes while monitoring electroencephalograms (EEGs) as they listened to a series of statements. The statements were recorded by actors told to come across as either confident, nearly confident, unconfident, or neutral. The researchers identified positive peaks in brain activity in all of the volunteers' brains after playing about 200 milliseconds of the voice recordings, regardless of confidence level, but confident speech sparked significantly higher peaks of brain activity than unconfident speech. Nearly-confident speech triggered additional brain activity after another 130 milliseconds elapsed, suggesting that it required slightly more time for decoding. What these results indicate is that confident speech grabs the most attention and demands the highest level of processing speed from the brains of listeners. And the effect is almost instantaneous; our brains are able to determine whose voice merits the most consideration before we've even considered what's being said. To put this in context, it's correct to say that your brain figures out whether someone's voice is worth hearing faster than a blink of an eye (which requires between 300 and 400 milliseconds).[8]

NEGOTIATING FAIRNESS: INDIGNITY VERSUS INTEGRITY

Like the monkeys in Dr. Santos's lab, humans are endlessly engaged in a tricky game of give-and-take. Dr. Santos coined the term *monkey-nomics* to describe this interplay among the capuchins. When a monkey feels wronged in a negotiation, he or she will refuse to participate any longer, or at least until the wrong has been rectified (which usually means that the other monkey hands over the grapes). Humans face similar circumstances all of the time, albeit with additional layers of complexity that make knowing when to draw the proverbial line all the more challenging.

Let's say, for example, that you are negotiating with someone about how to split a sum of money that you both can rightly claim, but that the other person, regrettably for you, has in his possession. You have one opportunity to make the deal, and the other person is not obligated to keep negotiating with you after this. Since he has the upper hand, the person you're negotiating with says that he thinks a 70–30 split is fair, with 70 percent going to him. If you accept his terms, you get 30 percent of the money. If you reject his terms, you get 0 percent. You believe the terms to be unfair, but if it's the difference between 30 percent and nothing, you'll take the 30 percent, right?

Maybe not. Instead, you might reject the offer as a symbolic way of expressing your anger and take the opportunity to tell the unfair dealer exactly what you think of him, money be damned.

OK. But now imagine that you are negotiating with someone who has been informed that she can unilaterally decide how much of the money to give you and you have no say in the outcome. In other words, as far as she's concerned, she can dictate the amount and she doesn't care what you decide—in fact, she'll never even know. On your side of the deal, however, all you know is that you are going to be offered a sum of money just as you were in the first deal, and you can choose to reject or accept it. You cannot, however, discuss the deal with the other person and voice whether you believe the deal to be fair or unfair—you have no recourse.

So once again you are offered 30 percent of the money, and this time not only are you faced with 30 percent or nothing, but you're also denied the satisfaction of telling off the unfair dealer or even symbolically protesting. This time it seems clear—you take the money, right?

Once again, quite possibly not. But why not? You have no chance of trying to make the deal fairer, and no opportunity to express your disgust, so what's making you still turn down the money? That's precisely what a study published in the *Proceedings of the National Academy of Sciences* investigated.[9] Participants were made to play both of the game scenarios above; the first is called the Impunity Game, a variation of the Ultimatum Game. In the Ultimatum Game, a proposer is given a sum of money and told to negotiate with a responder on how to split the amount. The responder has two options: (1) accept the amount proposed and both parties get the agreed-upon amount of money, or (2) reject the amount proposed, and neither party gets any money. The typical result of this game is that most unfair offers are rejected and the parties commonly agree to a 50–50 split.

In the Impunity Game, the responder can still reject the offer, but by doing so also loses any claim to the money. It's a "take X percent or nothing" deal. The typical results of this game are that between 30 and 40 percent of responders reject the offer in a show of symbolic punishment against the unfair party. The responder forfeits the cash but still says her piece.

The final variation is called the Private Impunity Game (the second imaginary scenario I gave you), in which the proposer is told that he can simply dictate the amount to be given to the responder. The responder, however, is told that she can still reject the offer but the proposer will never know what decision she made. In this case, the predicted result is that nearly all participants will act rationally and take the money, since they have no chance of recourse and no chance to make the proposer aware of their disgust.

That's the prediction, but surprisingly it turns out not to be true. The rejection rate of unfair offers is still a hefty 30–40 percent. The reason suggested by this study is, in a word, emotion. When faced with an unfair offer, we have the choice of rationally accepting the immediate incentive and ending the dispute or allowing an emotional response to dominate. We respond emotionally to unfair treatment for the same reason that a bear charges someone intruding on its territory. Because we know that a bear will act aggressively if it feels challenged, we avoid bears. The same dynamic applies to us: If someone is known to emotionally respond with anger and moral outrage to unfair treatment, he develops a reputation as someone to avoid crossing.

What this study also tells us is that not only are we concerned with consistency in our external reputations, but we're just as much, if not more, concerned with internal consistency. Our emotional response guards against accepting immediate incentives that compromise our integrity. Over time, this internal consistency that preserves integrity may also spruce up our external reputation. Simply put: Most of us would rather be seen as bears than sheep.

But here's a question: If it is really in our best interest to take less when the alternative is nothing, and stubbornly we still don't, have we really made the best decision? The tricky part is knowing what constitutes the "best" decision, and if perhaps taking a short-term hit to one's dignity is better for one's long-term interests. Unfortunately for us, our brains are not terribly efficient at making these judgments in short order. Time is always a factor. If the defensive emotional response is strong enough, it's likely no amount of calculating is going to thwart it, especially when things are happening fast (and they usually are).

Again, there is nothing black-and-white about the tendencies of a happy brain. Opinions on the scenario I just described will vary widely. But the main point remains the same: How our brains react in a given situation may very well undermine our best interests in the short and/or long term.

Now we will shift to another social dynamic that sits at the hub of every relationship (and not only those with other people): the power of influence.

Chapter 10

THE GREAT TRUTH RUB-OFF

"For certainly, at the level of social life, what is called the adjustment of man to his environment takes place through the medium of fictions."
—WALTER LIPPMAN, *PUBLIC OPINION*

OPINION BY COMMITTEE

Just last night you went to see a movie that defies simple categorization. Some films are plainly bad, others are obviously good, but this one doesn't filter so easily through the quality sieve. You find yourself searching the Web for reviews. When you stop to think about it, this is an odd thing to do *after* you have seen the movie. Normally you would consult reviews to find out if a movie is worth your time and money—but in this case you are doing it to find out what others thought about the film even though you have already watched it. You also post a note on Facebook, asking if anyone else has watched the movie and what they thought about it. You do the same with an email to a handful of friends.

Why are you pursuing these opinions? Why do they mean so much to you now that you've already seen the movie? Asked another way—what is lacking in your self-perceived ability to evaluate the film that you think others can supplement?

These questions flirt with uncomfortable psychological territory. We like to think of ourselves as self-sufficient arbiters of our experiences. To admit otherwise is to suggest that we are not independent thinkers. But for the last several decades, psychology—and, more recently, neuroscience research—has been providing evidence that "independent thought" is certainly not absolute, and possibly a figment of our egos' making. The truth is that our brains are not wired for complete independence. We are instead an exceptionally social species wired for *interdependence*. Ours is an existence of influence and counterinfluence—and none of us live on one-way streets.

Knowing this, it is much easier to understand why we would search out opinions on a movie, or anything else that challenges our sense of self-sufficiency. But this, like all tendencies of the happy brain, can quickly go too far. Without checking ourselves, continual reliance on others to form our opinions and make our decisions is damaging, principally because it prevents us from "reaching"—taking psychological risks that are important to the formation of character and strengthening of personality. There are, of course, also good reasons for this tendency, and in some cases they yield distinct benefits. The key is balance, as we'll explore.

YOU DECIDE, I DECIDE, YOU DECIDE

Rather than label the tendency to seek opinion reinforcement as a weakness of character, or another pejorative, we are better served to find out what drives it. Neuroscience research has been delving deeper to glimpse a few answers, with some intriguing results. For example, a study conducted by a team of researchers from Emory University in Atlanta wanted to find out what happens in the brain when "offloading" decisions to external sources; in the case of this study, the sources were financial experts.[1]

Study participants were asked to make financial choices both inside and outside a functional magnetic resonance imaging (fMRI) scanner. The choices were divided into two categories: "sure win" and "lottery." During the scanner session, researchers introduced a financial expert to the study participants and provided the expert's credentials to enhance

THE GREAT TRUTH RUB-OFF

his influence. The expert's advice was presented to participants on a computer screen above their financial-choice options. If the expert recommended an option, the word *Accept* was displayed above it; if he advised against the option, the word *Reject* was displayed. During half the trials, the word *Unavailable* was displayed, indicating that the expert had no advice for that decision.

The results indicated that both behavior and neural activation patterns were significantly affected by expert advice. When given an *Accept* signal by the expert, participants tended to make decisions based on the advice. Simultaneously, neural activity correlating with valuation was witnessed in the absence of expert advice; no significant neural correlations with valuation were seen in the presence of expert advice. In other words, the brain appears to offload the burden of figuring out the best decision when given expert financial advice. When the expert's advice was available, the participants' brains simply did not have to work as hard, so they didn't.

When this research was first published, it was covered by several media outlets with headlines like STUDY SUGGESTS THAT EXPERT ADVICE CAUSES OUR BRAINS TO SHUT OFF. In fact, the study did not suggest this at all. Much the opposite, it showed an active—not passive—tendency of a happy brain: conserve resources when credible external resources are available. It also showed that time is a critical factor affecting the offloading tendency. Study participants were given an average of 3.5 seconds to make a decision, which means that they did not have time to deliberate. The researchers intentionally designed the study this way to "push" participants' brains to make a decision. With less time, the brain must work even harder to calculate outcomes. Consider the energy spent to sprint a mile versus jog the same distance; more ground is covered in less time, but more energy is required to get there. When an external source becomes available to draw on instead of burning through internal resources, the brain is happy to accept. The brain imaging results of this study show an "attenuation" of neural activity correlating with valuation—that is, a tapering or reduction of activity—which is exactly what we would expect to see from a brain offloading the resource burn to someone else. Another way to think of this is to envision a racecar driver trying to get an edge in the race without having to expend more energy. Burning more fuel will result in more pit stops, so instead the driver "drafts" another car, effectively pulling energy away from that car to supplement his momentum. Likewise, the brain drafts external sources, thereby conserving its own.

PEER POWER PLUS

Unquestionably, peers exert a great deal of influence on each other even aside from our brain's crafty energy-conservation tendencies. Adolescence is nothing if not a working model of peer influence in its purest form. But psychologists have wrestled with a tough issue when trying to isolate exactly what drives the juggernaut of peer influence: Is it simply the desire to look "cooler" (for show), or does peer influence actually change our minds? A study by a team of psychologists at Harvard University used a combination of social psychology and neuroscience methodologies to find out if peer influence really can change how people value something; in this case, the attractiveness of a face.[2]

Fourteen male participants rated pictures of 180 women's faces on a computer monitor. For the majority of the faces—after they'd made their own rating—the students were shown the average rating given to that face by hundreds of previous participants. Unknown to the participants, the researchers had made up these ratings, which were sometimes higher than the participants' own rating and sometimes lower. Later, the participants rated the same faces again while undergoing a brain scan. The results indicated that viewing the faces had a definite effect on reward-related regions in the participants' brains—and that this effect depended on the feedback the participants had received earlier about how their peers had rated those faces. In other words, the peer feedback genuinely changed the participants' attitudes about facial attrac-

THE GREAT TRUTH RUB-OFF

tiveness. This held true even for faces participants had rated as equally attractive. If participants were told that those faces had been rated as more attractive by previous participants, greater reward-related brain activity was observed and they also increased their ratings. In contrast, the faces they had earlier been told were rated as less attractive by peers triggered less reward activity and were now rated as less attractive by the participants. The reason for these effects seems, again, to be a matter of neural patterns. The researchers who conducted this study believe that the same neural structures that guide us toward highly valued outcomes across a range of things—including food, water, and reproduction—are at play when we conform to others' opinions. The brain interprets the different value ratings and opinions of others as a signal that adjustment is needed to more effectively target the best outcomes. Hence, "If it's good for them, perhaps it will be good for me."

SENDING IDENTITY SMOKE SIGNALS

Another aspect of the influence effect has much to do with *who* is choosing *what*.

As a social animal, we have a deeply rooted desire to belong to a social group—a preferred tribe, if you will. When members of a given social group use or approve of something, it sends a signal to others that the thing in question is good for the group—that it is consistent with the tribe's identity. Researchers studying this dynamic use the terms *conformance* and *convergence* when separating out whether or not influence is identity based. As it turns out, we can reliably predict whether someone will be influenced in, for instance, a purchasing decision. We are more likely to conform our behavior to the groups' when we witness others buying a practical product like toothpaste, for example, because the group-purchase signals that this product is superior to the rest. In that case, it doesn't matter if the others are or are not part of a particular social group—it is the choice of the group overall that matters.

On the other hand, we are more likely to converge with (or simply join) those of an esteemed social group when we witness them buying high-ticket items—items that signify the group's status. In this case, it definitely matters who is doing the buying. If Ron's social group is keen on Mercedes-Benz cars, he's probably not going to go out and buy a Lexus (unless there's a distinct social advantage in doing so).

The first example illustrates choice conformity that is not identity based. The second example illustrates choice convergence, which is tightly wrapped up with identity. In both cases, choice was influenced by others, but conformity is rooted in seeking highest value; convergence is about *who we are or think we should be.* The interesting thing about identity-based decisions is that we have a harder time explaining why we made them, though we will come up with multiple reasons that are tangential to the core reason ("sleek styling, smooth shifting, impressive handling," etc.). Rarely will someone reply, "This car most closely fits the identity of my social group, and since my identity is derived in part from the group's identity, I bought it."

BEWARE: TRUTH MIRAGES ABOUND

So, drafting external resources to process decisions and opinions is not necessarily bad; in fact, it's a crafty conservation strategy that often serves us well. And peer influence also appears to be a function of our brains' adaptive strategy to locate high-value resources and also to signal our affinity with a social group. That is all well and good (much of the time), but the flip side is that these tendencies can also predispose us to influence by propaganda and cons of every stripe.

Two dynamics fuel the brain's acceptance of these influences: repetition and something psychologists call *cognitive fluency.* Since the early days of studying propaganda used during World War II, psychology research has demonstrated that the more a message is repeated, the more likely we are to believe it—particularly if we are paying *little* attention. Counterintuitive as it may sound, the series of glancing blows from oft-repeated messages is what eventually locks us into the "illusion of truth." The more focused we are on the message, the less likely we are to be influenced.

Cognitive fluency refers to our brains' tendency to accept messages that are easy to understand and effortlessly fit into existing schemata (referring back to chapter 1)—and, when positively employed, it is a skill crucial to learning.[3] The reason that persuasive messages are short, pithy, and digestible in seconds is that we process them so quickly that they become familiar without us even noticing. While conventional wisdom holds that familiarity breeds contempt, in the world of influence, familiarity breeds acceptance. This is again in large part due to the brain's proneness to conserve resources: Familiar messages require fewer resources to decode and process, and a happy brain is happy to take the less strenuous route.

Conversely, messages that are harder to process elicit the opposite effect: We are less likely to believe them. For abundant anecdotal proof of this, consider the difficulty policymakers have attempting to explain complicated issues to the public (any science-based issue faces this problem), and how equally hard it is convince us that the more complex position merits our belief, when a sea of simpler and more influential messages washes over us every day. The problem facing anyone trying to communicate complex messages is that our brains are not natively inclined to tackle such messages. Substantial effort is required to force one's attention toward less simple, less pithy messages when so many low-resource vittles are out there to digest with negligible attention required.

Advertisers and political strategists know this, of course, and capitalize on the fact that fostering an illusion of truth is what generally wins over voters and consumers.[4] It is hardly an exaggeration to say that almost every political campaign waged in the United States is a battle of illusions of truth. The more money a candidate has at her or his disposal to craft more effective messaging strategies, the more likely she or he is to win.

If you cannot get your illusion of truth out there enough times to counteract your opponents' messages and draw in constituents, your efforts are handicapped—likely beyond repair. It's a tragically simple algorithm that has dominated the political and consumer marketplace for well over a century.

In more recent elections, we've seen this dynamic taken to an extreme with blatantly fraudulent "news" repeated through social media, wielding a scary amount of influence on voters. In this case, the illusion of truth covers boldface lies that are manufactured to appear true to people seeking evidence to confirm their positions (referring back to *confirmation bias* in chapter 1).

It is worth noting that research has also been conducted to find out just how many times a message should be repeated for optimal effect. These studies suggest that we invest the most confidence in a message when it has been repeated three to five times. When we are saturated beyond that point, repetition loses its persuasiveness and may even reverse the effect altogether.

ALL ABOARD THE NARRATIVE TRANSPORT EXPRESS!

Needless to say, our brains' internal governor for accepting or rejecting external influence isn't always at the top of its game. You may be surprised to find out, for example, that even the fictional characters in our favorite television shows can persuade us to alter our thinking on touchy topics. That was the conclusion of a study showing that organ donation, when depicted favorably in popular television dramas, gets a boost in the public sphere. This might be good or bad, depending on how you look at it.

For some time now, research has been showing that television is a potent way to facilitate what psychologists call *social learning*—the tendency of people to model attitudes and behaviors of others under particular conditions. Two conditions are requisite: attention and memory.

Engaging television dramas that draw the viewer in to their narratives meet both conditions—they absorb attention and catalyze memory formation. When a viewer strongly identifies with a particular character in the drama, the effect—referred to in psych circles as *narrative transport*—is even more potent. In this case, a research team led by Lauren Movius, professor of psychology at Purdue University, wanted to know if depictions of organ donation in TV dramas like *CSI*, *Numb3rs*, *Grey's Anatomy*, and *House* would influence learning about organ donation and increase motivation to become a donor. They also wanted to know how, or if, accuracy of the information influences learning and motivation.[5]

Participants were asked to watch a selection of episodes from popular TV dramas with story lines that included both positive and negative depictions of organ donation, and then complete surveys that assessed a range of factors related to how strongly the viewer had been influenced by the story lines (and no small potatoes here; more than five thousand people completed the *House* survey).

The results indicated that viewers who were not organ donors before watching the dramas were more likely to decide to become one if organ donation was portrayed positively and if characters in the show explicitly encouraged it. Viewers who reported emotional involvement with the narrative were significantly more likely to become organ donors. And, finally, viewers clearly acquired knowledge from the content of each drama—whether or not it was accurate.

And that's the "depending on how you look at it" part of this. The study is really telling us a couple of different things: Emotional involvement with narrative affects the way people think and supplies knowledge that may very well not be true. Most people would probably agree that organ donation is a social good, and if TV dramas encourage it then all the better—but, the troubling part is that the same dynamic driving the good can also serve up the bad with equal effectiveness. Pseudoscience, anti-vaccine alarmism, and quackery of every flavor spreads just this way.

In another study, researchers wanted to find out if this effect holds similarly true on the big screen—this time with smoking as the modeled behavior. If a viewer strongly identifies with a particular protagonist in a movie, will that protagonist's smoking influence the viewer's thoughts about smoking?

Turns out, it does. Greater identification with a smoking protagonist predicted: (1) stronger "implicit associations" between the viewer and smoking for smokers and nonsmokers (in other words, associations that they were unaware of or wouldn't explicitly admit to), and (2) increased desire for those who already smoke to go light one up.

The researchers concluded that when we watch a movie, we often identify more with one character for any number of reasons. Our attention is engaged by his or her emotions and behaviors and we slide into the character's cinematic shoes, and, just as in real life, it's easier to be influenced by someone whose shoes we're trying on. The stronger this identification becomes, the more our thoughts are influenced, with behavior often following suit.

THE SWAY OF METAPHOR

Whether in person or on screen, one of the strongest influences on our thinking is woven into the verbiage all of us use in discussions big and small: metaphors. Let's say that we are comparing cities we have visited or would like to visit, and I mention one that I have not yet been to but you have. You say, "It's a massive, stinking cesspool filled with garbage and crawling with every form of filth imaginable." Immediately my mind conjures an image of a filthy retention pond covered with scum, loaded with trash, and lousy with rats and roaches. How close the metaphor you have chosen is to actually describing the city is debatable, but in the few minutes we are speaking, this doesn't really matter. What matters is that you have provided the metaphorical rudiments for me to construct an image that is now schematically associated with the city in my mind. One day I may visit that city and determine that your metaphor was inaccurate, or I may conclude that it was spot-on. Until then—or until I come across information that contradicts or verifies your description—the image will be there. And even after that, I'll find removing that image from my mind very difficult.

That is the power of metaphor—a power so subtle we barely notice how much it impacts our thinking. Researchers Paul Thibodeau and Lera Boroditsky from Stanford University demonstrated how influential metaphors can be through a series of five experiments designed to tease apart the "why" and "when" of a metaphor's power.[6] First, the researchers asked 482 students to read one of two reports about crime in the city of Addison. Later, they had to suggest solutions for the problem. In the first report, crime was described as a "wild beast preying on the city" and "lurking in neighborhoods." After reading these words, 75 percent of the students put forward solutions that involved enforcement or punishment, such as building more jails or even calling in the military for help. Only 25 percent suggested social reforms, such as fixing the economy, improving education, or providing better healthcare. The second report was exactly the same, except it described crime as a "virus infecting the city" and "plaguing" communities. After reading this version, only 56 percent opted for greater law enforcement, while 44 percent suggested social reforms.

Interestingly, very few of the participants realized how affected they were by the differing crime metaphors. When Thibodeau and Boroditsky asked the participants to identify which parts of the text had most influenced their decisions, the vast majority pointed to the crime statistics, not the language. Only 3 percent identified the metaphors as culprits. The researchers confirmed their results with more experiments that used the same reports without the vivid words. Even though they described crime as a beast or virus only once, they found the same trend as before. The researchers also discovered that the words themselves do not wield much influence without the right context. When Thibodeau and Boroditsky asked participants to come up with synonyms for either "beast" or "virus" *before* reading identical crime reports, they provided similar solutions for solving the city's problems. In other words, the metaphors only worked if they framed the story. If, however, they appeared at the *end* of the report, they didn't have any discernable effect.

How Language Shapes Our Perception

Shakespeare wrote, "What's in a name? That which we call a rose by any other name would smell as sweet."[7] According to Stanford University psychology professor Lera Boroditsky, that's not necessarily so. Focusing on the grammatical gender differences between German and Spanish, Boroditsky's work indicates that the gender our language assigns to a given noun influences us to subconsciously give that noun characteristics of the grammatical gender. Take the word *bridge*. In German, *bridge* (die brucke) is a feminine noun; in Spanish, *bridge* (el puente) is a masculine noun. Boroditsky found that when asked to describe a bridge, native German speakers used words like *beautiful, elegant, slender*. When native Spanish speakers were asked the same question, they used words like *strong, sturdy, towering*. This worked the other way around as well. The word *key* is masculine in German and feminine in Spanish. When asked to describe a key, native German speakers used words like *jagged, heavy, hard, metal*. Spanish speakers used words like *intricate, golden, lovely*. Boroditsky even created her own language (called Gumbuzi), with its own feminine and masculine grammar assignments, to test the hypothesis from scratch. After only one day of learning the new language, participants began using descriptions of nouns influenced by grammatical gender. Boroditsky's work suggests that how we see the world is strongly influenced by the grammar we internalize from an early age.[8]

JUST HOW MALLEABLE ARE WE?

If reading this chapter has so far left you unconvinced that our brains can be cajoled and swayed by any number of influences, read on—this next discussion might be the clincher. What if I were to tell you that your judgments about "good" and "bad" are heavily influenced by whether you are right-or left-handed? This is actually not a new finding; past studies have demonstrated that we are, in fact, prone to make judgments that correspond with the side we act more fluently on. Right-handed people prefer products and people on their right, left-handed people prefer the left. The question is, why? Are these preferences hardwired in our brains, or are they learned over time? Researchers Daniel Casasanto and Evangelia Chrysikou from the New School of Social Research and University of Pennsylvania investigated this question using a series of experiments that tested both possibilities.[9]

To address the first question (are these tendencies hardwired in our brains?), the researchers recruited thirteen right-handed patients who had suffered cerebral injuries that weakened or paralyzed one side of their bodies. Five remained right-handed. The rest lost dominant control of their right side and became effectively left-handed. The patients were shown a cartoon of a character's head between two empty boxes and told that he loves zebras and thinks they are good, but hates pandas and thinks they're bad (or vice versa). Then they were asked to say which animal they preferred and which box, left or right, they'd put it in. All of the patients who were still right-handed put the "good" animal in the right box. All but one of the new lefties put it in the left. So it would seem from this experiment that the brain's preferences (right or left) can definitely be altered—but is the change due to a neural wiring adjustment or to new learning? To rule out neural wiring as the answer, the researchers then took a group of fifty-six healthy right-handed people and asked half to wear a ski glove on the left hand and half on the right. The participants were instructed to pull dominos from a box, two at a time, using one hand for each, and place them symmetrically on dots spaced across a table. If a domino fell, they were to set it upright with the appropriate hand only— in other words, half of the right-handers were turned into left-handers for the duration of the experiment. They were then escorted to another room and administered the same animal-box task as the brain-injured patients. The results: three-quarters of those with ungloved right hands put the good animal in the right box; two-thirds of the temporary lefties put the good animal in the left. It took all of twelve minutes' worth of "training" to change a significant percentage of right-hander's loyalties to the left. What this and similar studies tell us is that even very basic judgments that we make every day are influenced by factors as innocuous as which hand we use the most. Does that mean we cannot trust ourselves to make sound, rational decisions? No. But it does mean that more of our decisions, opinions, and judgments are affected by a much greater range of influences than we know. And, as we are about to see, it also means that our brains are psychosocial germ mongers.

Chapter 11

HOW YOUR BRAIN CATCHES PSYCHOSOCIAL COLDS

"It would be difficult to exaggerate the degree to which we are influenced by those we influence."

—ERIC HOFFER,
THE PASSIONATE STATE OF MIND

YOU FEELING ME?

You are at a party with a few friends and everyone appears to be having a good time. The chitchat is casual and upbeat, as you would expect. The music is loud, and it's hard to hear anyone talking more than a couple of feet away, but you begin hearing what you think is yelling coming from the far side of the room. You tune in more, and now you're sure it's yelling—two male voices shouting louder and louder over the music. You start walking that way, along with several other people, and soon everyone else at the party except for the two men yelling are circling around the scene. Someone shuts off the music—now full focus is on the ruckus. It isn't long before one of the guys throws a punch. The other guy avoids it and tackles the first guy, who is elbowing the back of his opponent's head while both of them fall backward onto an end table that collapses upon impact. Eventually they are pulled apart and told to leave—but everyone knows what is going to happen when they exit, so the mass of people trail behind the two as the fight resumes outside.

Now the crowd is yelling—some people for one of the guys and the rest for the other. Then a couple of those people begin shouting at each other, followed by a few more, and soon the formerly calm and friendly crowd is a bomb on the brink of detonating. All of this happens in the span of about six minutes. Party over.

That cheery tale is an overt example of the psychosocial contagion phenomenon—the tendency to become "infected" by the emotions, thoughts, and behaviors of others. A happy brain catches emotional contagions rather easily and is also happy to pass along the bug. This, as with all tendencies of a happy brain, is both good and bad. Anger (or, in the party example, anger mixed with hysteria) is just one of several contagions psychology research has identified. Here are a few more:

Blame
Stress
Fear
Disgust
Anxiety
Apathy
Happiness
Moral outrage
Risk perception
Binge eating
Unethical behavior

One of the largest studies to date on emotional contagions focused on how happiness spreads through large social networks. The study used twenty years of data from the massive Framingham Heart Study to identify several important characteristics of contagious happiness.[1] First, the study suggests that happiness spreads over three degrees of separation, such as friends of a friend's friend. Researchers also found that people who are surrounded by happy people in the near term have a significantly greater likelihood of future happiness. And they found that happiness is a potent force for gluing people together, regardless of whether those people were similar to begin with. In short, happiness is highly contagious and its effects are lasting—unless or until geographic separation gets in the way (the one thing that cuts the infection short).

Your Sweat Makes Me Feel Risky

People are obsessed with managing their sweat because we think it's embarrassing in social situations (the dreaded underarm pancakes!). But a well-crafted study suggests that there's far more to our sweat than meets the eye; indeed, the sweat of others may be influencing us in ways we don't realize. Researchers collected sweat samples from people who completed a high-rope obstacle course and placed the samples in odorless tea bags, which were then placed under the noses of people about to gamble. Other gamblers were outfitted with sweat samples from people who had just finished riding an exercise bike. Gamblers sniffing the high-ropers' sweat took longer to make decisions, but eventually took significantly larger gambling risks compared to the bike-sweat-sniffing gamblers. Since there was no difference in how the sweat in either group smelled (everyone said the teabags smelled equally horrible), it appears that anxiety-laced sweat triggers riskier behavior than normal sweat. No one is quite sure why this is the case, but since the animal world is full of chemical-influence examples (think of ants and bees, for instance), it's not hard to believe that humans also send signals in ways that seemingly defy the senses.[2]

Additional behavioral-contagion findings from the same study include that we are 61 percent more likely to smoke if we have a direct relationship with a smoker. If your friend of a friend is a smoker, you are still 29 percent more likely to smoke. Even at a remote third degree of separation (friend of a friend's friend), you are 11 percent more likely. (The third-degree-of-separation findings have been challenged by some statisticians, but even without those findings, the overall impact of the research is powerful.)

A different study conducted by the same researchers, this time tracking social-network contagions over a thirty-two-year period, determined that if your spouse becomes obese, the odds of you becoming obese increase by 37 percent. And if a close friend becomes obese, the odds jump to 57 percent that you'll also pack on the pounds.[3]

We may reasonably debate the percentile accuracy of findings in studies like these, but the specific numbers aren't really the point. The findings describe patterns of influence that occur person-to-person, group-to-person, and group-to-group, in the short-term and over longer periods of time. We can't escape these patterns because they're integral to who we are.

SYNCHRONIZED ANXIETY

Psychosocial contagions spread because we humans are socially interdependent. We not only overtly influence each other but also spread influence without knowing it, through a form of emotional synchronization. One way to think about this dynamic is to imagine a flock of birds feeding on the ground until something startles a few of them and they take flight. Within seconds, the entire flock is taking off and flying in the same direction. The out-of-hand party scenario is one example of this. Another example is the way in which anxiety spreads through groups. In a group of any size, there will be some people more prone to anxiety than others. But research suggests that when the entire group is exposed to an anxiety-provoking stimulus, everyone eventually reaches the same level of anxiety no matter how emotionally controlled they were initially.

David Eilam, a researcher at Tel Aviv University, measured how groups of voles responded to threats produced by vole-hungry barn owls. Voles are a favorite rodent of researchers because they display very distinct social qualities, like humans, and because they display a range of anxiety responses (also like humans). When barn owls flew over the cages of individual voles, each of the animals' nervousness increased by about the same amount, as measured by standard behavioral tests. As a group, the animals continued to be all over the anxiety map. But when Eilam took groups of voles with different individual anxiety levels and exposed them to barn owls, the anxiety spread throughout the group and all of the voles became nervous wrecks.

Eilam believes that behavioral norms might be beneficial for social animals during a crisis. This convergence to similar behaviors may help explain why humans turn to religion and other rituals after a major catastrophe. These ceremonies, Eilam says, may keep the most anxious humans from going over the edge.[4]

BEWARE THE BLAME MONSTERS

Blame is an especially intriguing contagion, illustrated to near perfection in a *Twilight Zone* episode (circa 1959) titled "The Monsters Are Due on Maple Street." The setting is a calm suburban street at night that emotionally erupts after what appears to be a meteor flies overhead and the entire street experiences a total power failure. Then the neighbors receive shady news that humanlike aliens from the meteor have been spotted invading Maple Street. Soon strange things start happening, like the lights in one house turn on and off and then a random car on the street will start up with no one in it. It doesn't take long for the neighbors to begin accusing each other of being alien invaders. Blame spreads through the community until finally someone is mistaken for an alien and killed. All the while, two aliens sit on a nearby hill, controlling the power and marveling at how easy it is to manipulate human emotions.

Oversimplification aside (though masterfully told by *Twilight Zone* creator Rod Serling), the story effectively hits the point that blame spreads fast and usually ignites a few other dark emotions along the way. Science agrees. A study conducted by researchers from the University of Southern California and Stanford University suggests that blaming someone in public is the psychological equivalent of coughing swine flu into a crowd.[5] Over the course of multiple experiments, researchers showed that witnessing someone play the blame game significantly increased the chances of others' blaming someone else for their failures—even when those failures had nothing to do with what they witnessed. Blame contagion is essentially about self-image protection. The study authors believe that when someone watches another person level blame, the implicit takeaway is that self-image protection is a goal to which she should also aspire. In this study, blame became less contagious if people wrote down and affirmed their values before they witnessed someone attribute blame, which acted as a "blame antidote." The more self-affirmed people became (the more of the antidote they took), the less they felt the need to protect their image.

THE APATHY VIRUS

During every political season we hear about the power of apathy, a dull force that drains momentum and saps passion. But for all of the media coverage it receives, apathy is still a poorly understood condition. It's not merely the absence of commitment but also a potent psychosocial contagion that works in two directions.

Research has uncovered how apathy works by studying its effects across groups, spreading subtly but powerfully by way of indecision. The results of several studies can be summarized thusly: when we are not strongly committed to a goal, seeing apathy in others triggers and reinforces apathy in us. Conversely, when we are strongly committed to a goal, seeing apathy in others fuels our commitment.

The 2016 US presidential election provided the perfect case study for understanding apathy's contagious effects. The "lose-lose" feeling much of the population had over the year as the election neared kept goal commitment low. Millions of people said that they were undecided right up to the end, and millions more were influenced by the prevailing apathy to simply give up. The apathy contagion spread over many months and compounded in intensity with each new reason to feel like this suffocating election was a lost cause. At the same time, the resolve of those who were strongly committed one way or another actually strengthened as the apathy of others grew.

Let's throw this under a political strategist's microscope for a moment: if you want to capitalize on the power of apathy, the thing to do is amplify the apathy in the population, because by doing so you'll "infect" the uncommitted. The more successful you are in amplifying apathy, the more successful you'll be in expanding it. Further, the more apathetic the uncommitted become, the more committed the staunch supporters grow.

This hinges, of course, on whether it better serves your campaign to have higher or lower voter turnout, since growing the apathy pool dissuades people from bothering to stand in long lines. Conversely, it energizes your strongest supporters who see the apathy in others as more reason to stay the course. But that pool—the strongly committed—doesn't expand very much. It's the apathy pool that expands, and if it serves your objectives to keep people out of the fray, you'll want to make sure it expands as much as possible.

The most important factor in all of this is whether people are committed to a goal—a purpose for directed thought and action. Apathy is more difficult to manipulate as a contagion when commitment serves as a psychological buffer. But when commitment is fractured (as it surely was for a year leading up to the 2016 presidential election) those who stand to gain from manipulating the apathy contagion have much to work with.

CATCHING THE EMPATHY BUG

More and more research suggests that happy brains have difficulty differentiating between observing an action and actually participating in it. Empathy—a powerful contagion in its own right—seems to hinge in part on our ability to "take on" another's emotions through vicarious experience. I always think of this when watching a comedian fall flat. I can feel the embarrassment as if I'm standing there on stage, looking at a room full of blank stares. Something very similar occurs when we become infected with the emotions of others—as if our brains struggle with separating ourselves from what we see happening in those around us. Research has even found that our brains respond to the pain of those close to us as if we are in pain ourselves.

A study conducted by psychologists from Yale University and the University of California, Los Angeles, investigated this dynamic with an interesting angle: Researchers wanted to know if observing someone else exert self-control boosts or reduces one's own self-control.[6] Participants were asked to either take on the perspective of someone exerting self-control or merely read about someone exerting self-control. They were also asked to take on the perspective or read about someone *not* exerting self-control.

The results: Participants who *took on the perspective* of someone exerting self-control were unable to exercise as much self-control as those who merely read about someone exerting self-control. In other words, getting into the shoes of someone making the effort wore them out as if they were doing it themselves. On the flip side, participants who read about someone exerting self-control experienced a boost in their own self-control, compared to those who read about someone not exerting self-control. Reading resulted in a buttressing effect rather than a vicarious one.

The distinction between these results boils down to degree of psychological separation. Taking on perspective reduces psychological separation, and the more that gap closes, the greater the vicarious effect. Reading about something provides more of an opportunity to expand psychological separation (since the people you are reading about are not in front of you), which reduces the chances of vicarious effect.

The implications of these findings are quite practical. For instance, if a group of people is working on a project, and certain members are exerting an especially high degree of effort, this study suggests that other people in the group will experience a vicarious energy drain. An entire group's energy could be affected by the exertion of just one or two members. Another example is situations involving police officers, hospital staff, and other emergency workers, whose ability to maintain self-control is essential to their jobs. It's easy to see that if they experience vicarious depletion, anything from small breakdowns to catastrophic outcomes could result.

As an aside—this study also leads me to believe that "self-control" is at least half misnomer. Social influences affect it more than we know. On the other hand, regulating psychological distance—not something easily done—is a genuine application of self-control. If the pendulum swings too far in either direction, we either become wishy-washy emotional sponges or reinforced-concrete emotional silos.

WHAT YAWNING CHIMPS REVEAL ABOUT EMPATHY

Contagious yawning, I think you will agree, is no myth, and primate research is lately indicating that it may actually tell us something about the nature of empathy. Researchers at the Yerkes National Primate Research Center, Emory University, have been deconstructing the mechanism thought to underlie contagious yawning in both chimpanzees and humans.[7] They discovered that chimpanzees yawn more after watching familiar chimps yawn than after watching strangers yawn. Yerkes researchers Matthew Campbell and Frans de Waal (one of the world's leading primatologists) think that when yawning spreads between chimpanzees, it reflects an underlying empathy between those familiar with each other. They studied twenty-three adult chimpanzees that were housed in two separate groups. The chimps viewed several nine-second video clips of other chimpanzees, in both groups, either yawning or doing something else. They yawned 50 percent more frequently in response to seeing members of their group yawn compared to seeing others yawn.

Campbell and de Waal point out that yawns are contagious for the same reason that smiles, frowns, and other facial expressions are contagious: They are a measure of empathy, and—here's the fun part—empathy is *biased*. Odd as that may sound, what this research uncovers about chimps is just as true for humans—we empathize much more with those familiar to us, and this *familiarity bias* is demonstrated in something as basic as yawning.

As we have already discussed, the happy brain is quite the biased organ. And we would expect that to be the case, since brains evolved to ensure survival—and strangers to the tribe are more likely to be dangerous than those we know (unfortunately, it often turns out that those we know are also dangerous, but that's a different topic).

YOU CAN'T MIMIC THE TRUTH!

There is another significant drawback to catching the empathy bug, and we'll close out this chapter with a bit about it. Empathy research of the last twenty or so years has reinforced the mantra that mimicry—the tendency to imitate the behaviors and expressions of other people—not only smooths the wrinkles of social interaction but also facilitates better emotional understanding. The idea being that mimicry helps people feel what others are feeling and allows speakers to more accurately understand one another. And when it comes to truthful interaction, plenty of studies suggest this is the case. But what about during *deceptive* interaction? If mimicry helps me better understand you, will it also help me to know when you are lying?

A study conducted by psychologists at Leiden University in the Netherlands set out to answer exactly that question.[8] Participants were asked to interact with and mimic or not mimic people who claimed to have made a donation to charity—some of whom really had made a donation, others of whom were lying. In total, the experiment included three participant groups operating under three conditions: (1) told to mimic, (2) told not to mimic, (3) control—no instructions given.

To the Human Brain, Me Is We

A study from University of Virginia researchers supports a finding that's been gaining science-fueled momentum in recent years: the human brain is wired to connect with others so strongly that we experience what others experience, as if it's happening to us.

This would seem to be the neural basis for empathy—the ability to feel what others feel—but it goes even deeper than that. Results from the study suggest that our brains don't differentiate between what happens to someone emotionally close to us and ourselves, and also that we seem neurally incapable of generating anything close to that level of empathy for strangers.

To find this out, researchers had to get a bit medieval. They had participants undergo fMRI brain scans while threatening to give them electrical shocks, or to give shocks to a stranger or a friend.[9] Results showed that regions of the brain responsible for threat response—the anterior insula, putamen, and supramarginal gyrus—became active under threat of shock to the self; that much was expected. When researchers threatened to shock a stranger, those same brain regions showed virtually no activity. But when they threatened to shock a friend, the brain regions showed activity nearly identical to that displayed when the participant him- or herself was threatened.

"The correlation between self and friend was remarkably similar," said James Coan, a psychology professor in the University of Virginia College of Arts and Sciences who coauthored the study.[10] "The finding shows the brain's remarkable capacity to model self to others; that people close to us become a part of ourselves, and that is not just metaphor or poetry, it's very real. Literally we are under threat when a friend is under threat. But not so when a stranger is under threat."

Research in this category dovetails nicely with that conducted by evolutionary psychologist Robin Dunbar, whose work has shown that we seem to have evolved to cognitively connect in relatively small groups of roughly 150 or less people (often referred to as "Dunbar's Number"). Beyond that number, our brains strain to sync with others. From an evolutionary standpoint, this makes a lot of sense because chances of survival for ourselves and the group are amplified if we can devote the greatest level of cognitive resources to the task.

The results: Nonmimickers were significantly better at identifying the liars than mimickers—and this result held true when comparing the nonmimickers to mimickers and to the control group. Also worth noting is that all three groups were generally not very good at detecting lies (though the nonmimickers were the best), which buttresses another well-tested theory that, overall, people are just not very good lie detectors.

These results have several implications. That used-car salesman who is trying to put you into a "great deal"—be careful not to mimic his mannerisms. Ditto for just about any salesperson you come in contact with; while they may or may not be lying to you, it's best to put as much objective distance between you and them as you can, and mimicry reduces that distance. And that guy who shoulders up to you at a bookstore or coffee shop to tell you about a "great business opportunity"—well, don't even talk to that guy.

In his book, *Against Empathy*, psychologist and professor Paul Bloom makes a persuasive case that we overindulge in empathy at our peril. The source of the problem, argues Bloom, is that we chronically confuse empathy with compassion. It's possible to act compassionately without "taking on" the emotions of the other. As we've seen in this chapter, there's great value in preserving an emotional cushion between ourselves and those who could be recipients of our empathy, but that's easier said than done. That gap is filled faster than we realize, unless we consciously acknowledge its importance and act—however subtly—to preserve it. And we must always keep in mind that, unfortunately, there are plenty of people in the world who will gladly manipulate our empathy for their gain.[11]

Having said that, none of what we have discussed implies that we should not try to be empathetic when situations warrant doing so. Rather, in light of what we know about the potency of emotional contagions, it's not a bad idea to be aware of potential drawbacks of getting too emotionally enmeshed too quickly.

Chapter 12

THE HIDDEN POWER OF STUFF

"The things you own end up owning you."
—CHUCK PALAHNIUK, *FIGHT CLUB*

BRAINS AT MY FINGERTIPS

A few years ago I was part of a group that was making a presentation to state health agencies on effective ways to educate the public about air quality. During our last practice session before the real presentation, one of the seasoned and sly presenters brought in three massive bound documents and dropped them with a thud on the lectern. Before we started, I asked him what he was going to do with them. He replied, "You'll see."

When it was his turn to present, I did indeed see. Every time he made reference to research backing up his assertions, he lifted one of the documents high enough for the audience to see, and then judiciously dropped it onto the wood surface, just enough for everyone to feel the weight of it. I never asked him if the documents actually contained the research he was mentioning, but it really didn't matter. The effect was potent.

What I didn't know at the time is that this trick was making use of something cognitive psychologists call *embodied cognition*—the hypothesis that bodily perceptions, like touch, strongly influence how we think.

Another way to explain it is that our brain is not restricted to the space between our ears. Since our entire nervous system is integral to thinking, it makes sense that the physical sensations out in the world would influence our perception. What makes this hypothesis so interesting, however, is that these influences affect us without our notice. Let's take a look at a handful of experiments that illustrate the point.

HEAVY IS THE MIND

A study titled "Weight as an Embodiment of Importance," sheds light on the example I used at the beginning of this chapter. Over the course of multiple experiments, researchers investigated whether judgments of importance are tied to an experience of weight.[1] For a little context, consider how many ways in which weight—or facilitators of weight—overtly affect our judgments. In English, we use the term "weighty" to signify something substantial and important. We also use the term "gravitas" to connote seriousness, an elaboration on our understanding of gravity as a force exerting the power of weight over everything around us (and ourselves). We also think of weight as the arbiter of physical strength: The more someone can lift—or looks as if he or she can lift—the more impressive. Weight is even a socioeconomic force, as in the size of someone's car or SUV. I recall when the Hummer first arrived on the scene, we heard a lot about it being a "six-ton SUV," as if that specification made it more noteworthy than any other SUV.

In the study, a group of participants were first asked to estimate the value of several foreign currencies while they held a clipboard. Some held a light clipboard, others held a heavy one. As predicted, participants who held the heavy clipboards estimated the value of the currencies significantly higher than those who held light clipboards.

The second study repeated the first, but instead of judging currencies, participants were asked to judge the importance of having a voice in an important decision-making process (they were given a scenario involving a crucial decision affecting them being made by a university board). Again, participants holding heavy clipboards judged the importance of having a voice in the decision as more important than those holding light clipboards—a result showing that even something abstract, like making a decision, is tied to experience of weight.

In the final two studies, participants were asked to agree or disagree with arguments of varying strengths. This is a test of cognitive elaboration, one's tendency to assume and defend a strong position in light of given factors.

The results again showed that people holding heavy clipboards assumed stronger, more polarized positions than those holding light clipboards, and made significantly stronger arguments in defense of the positions. Opinions of those with the heavy clipboards were voiced more vituperatively than the others as well.

What makes this series of studies so impressive is that they cut across tangible and intangible variables (currencies versus decisions, arguments, etc.) and arrived at a quite consistent result: Experience of weight affects our thinking—and does so without our notice.

Bitter Taste, Bitter Judgment

If you've ever had a sip of the "bitters," your face probably scrunches up just thinking about it. According to a study from researchers at Brooklyn College, the horrible taste may do more than make you pucker. Researchers had fifty-seven undergraduate students rate their moral distaste for several morally dubious acts, like politicians taking bribes, two second cousins sleeping together, and a man eating his dog. Before they started rating the acts, the students drank shots of one of three drinks: Swedish bitters, sweet berry punch, or water. On a 100-point scale, with 100 being the worst rating for a morally reprehensible act, the students who drank the bitters gave the acts an average rating of 78; those who drank sweet berry punch gave an average of 60; and the water group gave an average of 62. The ratings of the punch and water groups were statistically the same, but the bitters group was significantly higher, indicating that the bad taste increased the students' moral disapproval.[2]

IF YOU'RE FEELING WARM AND FUZZY, IT MIGHT JUST BE THE COFFEE

If you have a falling out with someone and he starts ignoring you, he's "giving you the cold shoulder." If you feel emotionally close to someone, you have "warm feelings" toward that person. We're accustomed to using metaphorical language like this to describe human relationships, but do these words also imply more literal meanings?

A study in the journal *Psychological Science* delved into whether the actual experience of warmth or coldness influences our perception of social relationships. In other words, are temperature differences really tied to differences in social closeness and social distance?[3]

The study included three experiments; in the first, participants entered the lab and were handed either a cold or a warm beverage. They were then asked to fill out a questionnaire (which was just a prop for the study), and then asked to select a person they knew and rate their relationship with that person on a scale called the Inclusion of Other in Self, designed to determine the degree of closeness between the subject and the person he or she selected. At no time were the subjects made aware why they were holding a warm or cold beverage—all they knew is that they were being asked to complete a few questionnaires.

The results: Subjects holding the warm beverage had a significantly higher level of perceived closeness to the individual they selected than subjects holding the cold beverage, bearing out the hypothesis that physical warmth is tied to perception of social "warmth."

The second experiment investigated whether watching film clips in a warm or cold room influenced the choice of language used to describe the film, with the hypothesis being that warmer temperatures will influence subjects to use more concrete language (such as "John punched David") versus more abstract descriptions ("John is angry with David"). The results were that subjects watching in the warm room did in fact use more concrete language to describe the film than did subjects in the cold room, who used abstract terms to describe the same clips. Previous research has shown that use of concrete language strongly correlates with a sense of social closeness, whereas abstract language correlates with social distance.

Take Two Photos of Your Loved One and Call Me in the Morning

As everyone who has been hospitalized knows, the pain is easier to bear if you have loved ones nearby. A study investigated whether the pain relief we get from this kind of support can be achieved with a photograph of a supporter instead of the real thing. The subjects were twenty-eight women in long-term relationships. They were brought into a testing room and their partners were brought into another to have photos taken. The women underwent testing to determine their pain thresholds via thermal stimulation. Once the thresholds were established for each subject, they were then exposed to a series of conditions while experiencing pain, including (1) holding the hand of their partner as he sat behind a curtain, (2) holding a squeeze ball, (3) holding the hand of a stranger, (4) viewing a photograph of their partner on a computer screen, (5) viewing a photograph of a male stranger, and (6) viewing nothing. Subjects rated each condition's unpleasantness on a twenty-one-point numerical scale. Here's what happened: As expected, holding their partner's hand resulted in significantly reduced pain ratings when compared to holding an object or a stranger's hand. Viewing their partner's photograph also produced significant pain reduction when compared to the object and stranger conditions. Interestingly, viewing a photo was also marginally *more* effective than holding their partner's hand. What seems to be happening here is that our brains can be made to conjure mental associations with being loved and supported just by viewing a photo—and this effect is potent enough to actually reduce how much pain is felt. And, as the results suggest, in some cases a photo may be even more effective than the genuine article.

COME HEAVY AND SIT HARD

A well-publicized study published in 2010 did an especially nice job of bearing out the embodied cognition theory. Researchers from MIT, Harvard, and Yale performed six experiments exploring whether the hardness, weight, shape, and texture of certain objects affect our decisions about totally unrelated situations.[4] For example, the study shows that when you're negotiating a deal, it's better to sit in a hard, sturdy chair—doing so may lead you to negotiate harder than you otherwise would. And when you go for a job interview, be sure to carry your resume in a weighty, well-constructed padfolio; according to the study, job candidates appear more important when they are associated with heavy objects. And when you invite your date over for dinner, keep the setting "smooth"—objects with a rough texture make social interactions seem more difficult than they really are. So put away those glasses with the beveled edges and your evening will stand a better chance of success.

YOUR WAISTLINE CHANGES
HOW YOU SEE THE WORLD

As anyone who has ever tried knows, losing weight is hard. The more, the harder. It's not just that the process of dieting and exercising is difficult; the weight itself can feel insurmountable—particularly if you've tried and failed to lose it before. It's a big physical *and* perceptual challenge. But there's also a "hidden" challenge, uncovered by a study that probed another dimension of perception: our body weight changes how we see distance. Psychologist Jessica Witt, lead author on several studies examining the connection between our physical realities and perception, conducted the study.

"If you find yourself out hiking with a heavy backpack, hills are going to look steeper, distances are going to look farther, gaps across a river are going to look longer," said Witt when presenting the study findings at an American Association for the Advancement of Science (AAAS) meeting. "You're not seeing the world as it is, you're seeing the world in terms of your ability to act."[5]

Witt and her team conducted the experiment at a superstore, where they set up several traffic cones and asked people to estimate the distance from where they stood to the cones. The participants were broken out into a few weight categories, with most falling into a normal range, moderately overweight, or obese.

The results: The obese participants perceived the cones as farther away than participants in the normal or moderately overweight groups. The differences between the groups correlated with the degree of obesity; people carrying an additional two hundred pounds of body weight estimated twice the distance as the others.[6]

Those results suggest that obesity not only presents a physical health risk but also creates a perceptual obstacle that limits physical activity. This is because physical activities, such as walking the distance between the cones, seem even more strenuous than they really are; in turn, this contributes to restricted physical activity and, therefore, remaining obese.

The study builds on a raft of research into the hidden power of stuff, much of which has been conducted by Witt and her team, demonstrating how perception influences performance. Here are some of the findings:

- Hikers wearing heavy backpacks looking up at a hill perceive the slope as steeper than those with lighter backpacks (the same holds true for hikers who are tired versus well-rested).
- Golfers who sink more putts perceive the hole as larger than those who frequently miss.
- Softball players who hit well tend to overestimate the size of the ball; those who miss more often tend to underestimate it.
- Experienced parkour athletes (who climb and leap through and over city buildings and other structures) estimate wall sizes more

accurately than parkour novices, who tend to estimate walls as taller than they actually are.

- American football field goal kickers who tend to kick too low or too wide perceive the goal posts as taller or narrower than they actually are.

Having covered influence from multiple angles, we will now move on to topics so central to our lives, we would be lost without them. Literally, *lost*.

Part 5

MEMORY AND MODELING

Chapter 13

YOUR MIND IN REWRITES

"Time and memory are true artists; they remold reality nearer to the heart's desire."

—JOHN DEWEY,
RECONSTRUCTION IN PHILOSOPHY

ARE YOU SURE YOU SAW "X"?

As a consulting analyst for the US Environmental Protection Agency, I spent quite a bit of time conducting research to find out if a national campaign to raise awareness about childhood lead poisoning was reaching the eyes and ears of the right people: Parents of young children in cities with older housing. One of the ways the lead-awareness message was advertised was during previews at movie theaters in multiple cities. I conducted on-site research at theaters in Baltimore that included two types of in-person surveys: a pre-movie survey to determine peoples' knowledge of the lead-poisoning issue, and a post-movie survey to determine if people had been influenced by the lead-awareness message. The first survey established a benchmark of awareness; the second gauged how much the advertising had elevated awareness. At the same time, the post-movie survey determined to what extent people recalled the lead-awareness advertisement.

After the first couple of on-site studies, it became clear to me that something odd was going on. I knew going in that under the best circumstances, we would see at most a 30–40 percent recall rate for the advertisement (realistically closer to 25 percent). Yet we were getting close to 60 percent results at the movie theaters—double the expected rate, and virtually unprecedented for this sort of public outreach. So, in the next theater study, I decided we should begin asking a couple of new questions in the post-movie survey to determine whether people could recall details about the advertisement and not just a message. Doing this revealed that relatively few people could recall anything specific about the advertisement, although they claimed that they did recall seeing the lead-poisoning-awareness message.

What was really happening? After some digging, I reached a few conclusions—chief among them that people were unknowingly picking up on shards of information in the theater lobby (e.g., briefly seeing the EPA logo at a table we were using, hearing the words "lead poisoning" as they walked by other people who were being interviewed, or overhearing people who had completed the survey talking about it on their way out of the theater). The reality was that no more than 30 percent were actually recalling the pre-movie advertisement; others had assembled enough fragmented information that they were sure they *must have* seen it. The fragments comingled in their memory, and when they were asked a question that served as a sort of mental glue for the fragments, they were sure they had seen the ad.

This example illustrates a simple truth that we are typically reluctant to admit: Our memories are wrong at least as often as they are right. At best, they are incomplete, though we might swear otherwise. This affects countless aspects of our lives, and in many cases our memories—true or false—affect others' lives. But before exploring the fallibility of memory in greater detail, let's take a few minutes for an overview of what we know about how memory works.

A BRIEF PRIMER ON MEMORY[1]

Memory can be separated into two main divisions: explicit and implicit. Explicit memory (also called *declarative memory*) principally holds words, numbers, and events. Or, to use the parlance of neurobiology, it is memory that's *semantic* and *episodic*. When we are trying to remember what happened on the camping trip we took with our in-laws in late 2004, for example, explicit memory is engaged.

Implicit memory (also called *nondeclarative memory*) is where so-called muscle memory is found—the motor skills that, once learned, are always available to us. How is it that you never have to remember how to clip your nails or brush your teeth? Because implicit memory has you covered.

Those are the two main divisions of memory, but there are also two temporal (time-based) memory categories. The first is working memory, or what we usually call "short-term memory"—the category of memory that includes anything we are actively thinking about right now. In earlier models of memory, working and short-term memory were considered two different categories, with short-term thought of as a passive holding zone for information with closer links to attention and awareness. But more recently it's been suggested that these categories are synonymous.

The second category is long-term memory—the place where everything "remembered" resides. Long-term memory is engaged in the short term (e.g., word memorization) and in the long term (e.g., childhood memories). One way to think about short-term (working) memory versus long-term memory is that anything actively in your awareness right now involves short-term memory. In most views, short-term memory has extremely limited capacity—about seven items at a time. Long-term memory includes anything that is not currently active but could be recalled and made active. This could be anything from information you crammed for an exam yesterday to the address of the house you lived in when you were eight years old.

Perhaps the most exciting neuroscience discovery of the last several decades is that our brains are not static hunks of tissue but flexible and adaptive organs that change throughout our lives. The term used to describe this understanding is *brain plasticity*. The flexibility of your brain is essential to memory and indispensable to learning. Specifically, the "plastic" parts of our brain are synapses—the connection points that allow neurons to transmit signals between each other. Hypermagnification of synapses in an adult human brain shows synapses of various sizes and shapes, some are shaped like mushrooms; others, more like small hills; and others, like broad-based mountains. The incredible part is that in your brain and mine, synapses are morphing from one shape to the next depending on the need—how fluid the connection between neurons needs to be, for example—and this continues happening throughout our lives.

Another discovery related to brain plasticity is that the amount of activity between neurons corresponds directly to how strong their connection will continue to be. Cognitive scientists commonly use the phrase, "Neurons that fire together wire together." What this means in terms of memory is that the more intense the activity is between neurons constituting your memory of any given event, the more robust the memory will be. That is one reason why emotionally charged memories frequently percolate to consciousness in vivid detail. I can still remember almost exactly where I was standing outside my high school in Florida just after the space shuttle *Challenger* exploded. I was looking up at the sky and could see the entire shape of the explosion outlined in smoke. All "where were you when" moments have a degree of this same intensity, which makes them easier to recall than anything else that was happening at the time. In fact, I was writing the first edition of this book, history transpired: The president announced that Osama bin Laden had been killed. I feel as though I now have bookend memories of the horrible events of 9/11. I can recall in detail what I was doing when I first learned of the terrorist attacks in 2001, and I now have a memory of where I was when the mastermind of that attack was finally dispatched, almost ten years later. These are powerful, emotional events accompanied by intense neural activity. The imprint, though still imperfect, remains with us for a lifetime.

That is not true, however, for most memories. Even for the sharper memories born from strong emotions (often called *flashbulb memories*), time erodes the infrastructure, leaving cracks and gaps. Instead of remembering specific, explicitly accurate details, what constitutes memory over time are general impressions of events with spotty details—and the older we get, the spottier they become.

REMEMBER ONCE, FORGET TWICE

To figure out why that happens, among other perturbations, cognitive science has tackled memory more aggressively than perhaps any other topic. As a result, we have a growing wealth of research to draw upon to better understand the quixotic art of remembering. What we now know is that our brains happily reconstruct memories, though we are frequently fooled into thinking that the reconstructions are seamlessly recorded recollections.

What I want to convince you of in this chapter is that our memories are anything but concrete and can be altered with relative ease. That's the bad news. The good news is that imperfect memory is an evolutionary adaptation that serves our species well much of the time. Loss of memory, and creation of new memory, is central to a relatively efficient system of information processing that never sleeps. The selective movement of information into long-term memory is an adaptive marvel that allows our brains to store crucial pieces of information that we will rely on in the future, and shed information not worth holding onto. The process is not neat and tidy, and memory selectivity often works against us (think about how many memories you would love to forget). But when you view the process through the lens of species survival, it makes unassailable sense. In those terms, it is crucial to remember where the best sources of food are located; where the best hunting grounds are located; which areas to avoid lest you become something else's dinner; and how to return safely back home. For our ancestors, reliance on memory of particular details was a matter of life and death.

The problem for us moderns is that memory, incredible adaptation though it is, faces relentless challenges in societies driven by information. We simply have too much to remember at any given time, and the vast majority of our brains are not equipped to handle the deluge. Our expectations for what we *should* be able to recall are hardly in line with what our brains are capable of processing—which, by the way, is an enormous amount.

We have also adopted inaccurate metaphors for memory that lead us to incorrect conclusions. The "bookshelf" metaphor, for example, which suggests that when we need to recall a memory we simply find it categorized on a mental shelf, ready for consumption. Or the computer metaphor, which suggests that our brains store files on a cerebral hard drive that we can access as we would files on our laptop. These and similar metaphors are wrong for roughly the same reason: Memory does not reside in any one place in our brains, but rather is distributed across multiple brain regions. But because we more easily connect with a metaphor like those above than we do with the messy truth of distributed and reconstructed memory, the misunderstandings persist.

Next we will review a few experiments that put a finer point on just how changeable memory can be.

Think You Can Draw the Apple Logo from Memory?

Pick any ranking of publicly traded companies, and year after year you'll find Apple in the top three worldwide. To say that its famous logo is ubiquitous is an understatement. But right now, without any help, do you think you could draw the Apple logo from memory? If you're like participants in one memory study, you probably feel confident you could pull it off without peeking. But you may find, as they did, that it's much harder than you think.

Researchers from the UCLA Memory and Lifespan Cognition Lab conducted a handful of experiments to find out just how strong peoples' memories are of extremely common things, like the Apple logo—things they see every single day, multiple times a day.[2]

In the first experiment, they asked eighty-five undergraduate students (a mix of Apple and PC users) to draw the Apple logo from memory without any help. The researchers graded the drawings based on how well the participants positioned the leaf, the drew the shape and location of the bite, and reproduced the overall shape of the apple. They also asked the participants how confident they were that they'd get it right. The results: only one person drew the logo perfectly, and only seven drew it with just a few errors. The other seventy-seven people couldn't do it without major flubs, despite the fact that most of them said they were confident they could.

Another experiment replicated the first but this time asked participants to rate their confidence before and after the drawing test. Once again most rated themselves as quite confident they'd get it right, but their confidence dropped to just better than average after they finished. The reason—their poor performance (just as bad as the first group) woke them up to the fact that they really had no idea how to draw the thing.

The bottom line of these experiments is an affirmation of something that's hard to believe but consistently true: our visual memory of things we see all of the time is really quite poor. No matter how often we see something like a world-famous logo, there's rarely a reason for our brains to form detailed, long-term visual memories of it. And this is true despite our not-unreasonable confidence that we should remember it well. One reason for this is that our visual memories are handicapped by overexposure to commonplace images—quite likely because our brains have little reason to invest energy in memorizing something we see all of the time. As we've seen, our brains are misers when it comes to spending energy on everyday tasks. Instead of investing in top-to-bottom memorization of a common logo, the brain creates what the researchers call a "gist memory," which is little more than a shell of a memory that reminds us of what we're seeing when we see it, but provides little substance to re-create it from memory. Looking at it that way, it's no surprise that anyone thinks they could draw or recall the logo without peeking, because our brain fools us into believing we have a deeper visual grasp of the thing than we actually do. The truth is that we're likely to fail in memory tests like those in this study because normally it's unimportant that we remember more. Usually a gist memory gets us by just fine.

A PHOTO IS WORTH A THOUSAND WAYS
TO CHANGE YOUR MEMORY

Most of us realize that memory is fallible because of the little things that happen all of the time. We forget things like car keys, passwords, whether we turned off the oven, and so on. But how many of us would admit that our memory is susceptible to change from the outside? That's different from simply forgetting—something we all do on our own—because someone else changing our memory requires "getting in our heads," so to speak, right?

The truth is that this sort of outside-in influence does not take very much effort to accomplish—just a few images and a little time. A study conducted by Linda Henkel in the Department of Psychology at Fairfield University tested whether showing people photos of completed actions—such as a broken pencil or an opened envelope—could influence them to believe they'd done something they had not, particularly if they were shown the photos multiple times.[3] Participants were presented with a series of objects on a table, and for each object were asked to either perform an action or imagine performing an action (e.g., "crack the walnut"). One week later, the same participants were brought back and randomly presented with a series of photos on a computer screen, each of a completed action (e.g., a cracked walnut), either one, two, or three times. Other participants were not shown any photos.

One week later, they were brought back to complete a memory test in which they were presented with action phrases (e.g., "I cracked a walnut") and asked to answer whether they had performed the action, imagined performing it, or neither, and rate their confidence level for each answer on a scale of one to four. The results showed that the more times people were exposed to a photo of a completed action, the more often they thought they'd completed the action, even though they had really only imagined doing it. Those shown a photo of a completed action once were twice as likely to mistakenly think they'd completed the action than those not shown a photo at all. People shown a photo three times were almost three times as likely to think they had completed the action as those not shown a photo.

Two factors in this study speak to the malleability of memory. The first is time duration. The experiment was carried out with a week between each session, enough time for the specific objects and actions to become a little cloudy in memory, but not enough time to be completely forgotten. This lines up well with real-world situations, such as someone providing eyewitness testimony, in which several days if not weeks might elapse between recollections of events.

The second factor is repeat exposure to images. The study showed that even just one exposure to a photo of a completed action strongly influenced incorrect memory, while multiple exposures significantly increased the errors.

IF THE VIDEO SAYS SO, THEN I MUST BE GUILTY

So if static images can be used to manipulate our memories, what about video? After all, in a world dominated by endlessly pliable electronic media, you can never be 100 percent sure that what you're seeing on screen is what really happened. Two memory studies conducted by researchers Kimberley Wade, Sarah Green, and Robert Nash at the University of Warick (UK) illustrate that point nicely.[4]

In the first study, researchers wanted to know if they could convince people that they had committed an act they never did. To accomplish this, they created a computerized multiple-choice gambling task for participants to complete, which entailed increasing the winnings from a sum of money as much as possible by answering questions. The money was withdrawn from an online bank based on cues given to participants by the computer program: When they answered questions correctly, they were told to withdraw money from the bank; when they answered incorrectly, they were instructed to deposit money back into the bank. Subjects were videotaped while they completed the task.

Afterward, participants were asked to sit and discuss the task with a researcher. During the discussion, the researcher said he had identified "a problem" during the task, and then accused the participant of stealing money from the bank. Some of the participants were told that video evidence showed them taking the money (but they weren't actually shown the video), while others were shown video "proving" that they took the money. What the participants didn't know is that the video had been edited to make it appear as if they did something they had not. Participants were then asked to sign a confession stating that they did in fact take money from the bank when they should have deposited it back. Participants were given two chances to sign the confession, and by the end of the day, all of them did. In fact, 87 percent signed on the first request, and the remaining 13 percent signed on the second. Interestingly, even participants merely *told* that the video showed them taking the money eventually complied with the confession.

The next study used the same principle, but this time to see if people would accuse someone else of doing something they had not. Again a gambling task was used, but instead of one person completing it, two people placed side by side completed it—sitting not even a foot apart, with monitors in full view of each other. Subjects were videotaped as before, and the video was doctored as before to show one of the two participants taking money.

Afterward, the "innocent" participant was asked to discuss the task with a researcher, and told that video proof had been obtained showing that the other participant stole money. In order to pursue action against that person, the researcher said, the innocent participant would have to sign a witness statement corroborating the video evidence. Some of the participants were, as before, only told that the video existed, while others were shown the edited video (and there was also a control group neither told about nor shown video).

The results: When first asked to sign the witness statement against the other person, nearly 40 percent of the participants who watched the video complied. Another 10 percent signed when asked a second time. Only 10 percent of those who were only told about the video agreed to sign, and about 5 percent of the control group signed the statement. These results point to the alarming power of images to shape and distort memory—not only about others, but about ourselves as well. In the first study it wasn't only watching a video that made a difference; merely being told that a video existed made nearly as big an impact. And it is worth noting that in the second study, some of the people who signed the witness statement became so convinced that the other person was guilty that they went on to insert even more details of suspicious behavior, as if they knew the other person was doing something wrong all along.

TRUSTING YOUR WAY INTO FALSE MEMORIES

The examples we just discussed address what can happen to our memory when visual information is manipulated. Let's now remove the visual element and focus instead on how the integrity of information we receive affects memory—or, more precisely, the integrity of the information provider. If, for example, you follow a news commentator closely, reading everything he or she writes in whatever venue it appears, you may unknowingly be in a *trust trap*. Studies have shown that once we invest trust in a particular source of knowledge, we're less likely to scrutinize information from that source in the future. A study conducted by Elizabeth Loftus at the University of California, Irvine, and her team took this investigation a step further, showing that the trust trap can also result in the creation of false memories—and not only in the short term.[5]

Researchers crafted an experimental design in which they exposed two groups of participants to a series of images followed by narration about the images. The first group (referred to as the "treat-trick" group) received mostly accurate narration about the images. The comparison group received mostly misinformation. Both groups then completed tests of recall to determine how much accurate versus inaccurate information they remembered.

One month later, the participants were brought back to undergo the same experiment, except this time the treat-trick group was given misinformation during the narration (i.e., the "trick"), as was the comparison group. Both groups again completed tests of recall.

Here's what happened: In the first session, the treat-trick group had a significantly higher rate of true memory versus the comparison group (which we would expect since only the comparison group was given misinformation during this session)—at a rate of about 82 percent for the treat-trick group and 57 percent for the comparison group.

But in the second session, in which both groups were given misinformation one month later, the treat-trick group had significantly lower true memory recall than the comparison group: 47 percent versus 58 percent. The most likely reason for this effect is that the treat-trick group fell into a trust trap. Because information provided by the narrative source in the first session was accurate (and test scores were high as a result), participants believed the source to be credible and trustworthy. The comparison group, on the other hand, had no reason to invest trust in the original source and exhibited recall at roughly the same level for both sessions.

What's most interesting is the time frame of this effect. Researchers conducted the sessions a month apart, allowing ample time for a trust effect to wear off. But it didn't. Once again, we see that the real-world implications of this research. Eyewitness testimony can be changed when a witness listens to an information source she has previously trusted as credible (media, interrogators, or other people), and this study suggests that the window of opportunity for this effect is large. Any follow-up information received by an eyewitness from any number of sources can significantly alter his or her memory.

FALSE BELIEFS: SPAWN OF FALSE MEMORIES

If there's anything that cognitive-psychology studies have made clear over the years, it's that humans can be exceptionally gullible. With a little push, we're prone to developing false beliefs not only about others but also about ourselves with equal prowess—and the results can be, well, hard to believe. And at the core of many of these false beliefs live false memories.

For example, a study in 2001 asked participants to rate the plausibility of having witnessed demonic possession as children and their confidence that they had actually experienced one. Later, the same participants were given articles describing how commonly children witness demonic possessions, along with interviews with adults who claimed to have witnessed possessions as children. After reading the articles, participants not only rated demonic possession as more plausible than they'd previously said but also became more confident that they themselves had actually experienced demonic possession as children.

Another (less dramatic) study asked participants to rate the plausibility that they'd received barium enemas as children. As with the other study, participants were later presented with "credible" information about the prevalence of barium enemas among children, along with step-by-step procedures for administering an enema. And again, the participants rated the plausibility of having received a barium enema as children significantly higher than they had before.

A study conducted by researcher Stefanie Sharman at the University of South Wales sought to determine the effect of *prevalence information* (information that establishes how commonly an event happens, making it seem more likely and therefore more self-relevant) on the development of false beliefs.[6] Participants were asked to rate the plausibility of ten events from 1 ("not at all plausible") to 8 ("extremely plausible"), and how confident they were that they'd experienced each event from 1 ("definitely did not happen") to 8 ("definitely did happen"). The events included a range of the highly plausible ("I got lost in a shopping mall as a child") to the highly implausible ("I was abducted by a UFO").

How a Memory-Reinforcing Checklist Can Postpone Your Bucket List

A study published in the *New England Journal of Medicine* about the results of using surgery-safety checklists at major urban hospitals around the world underscores how dangerously fallible memory can be.[7] Around the globe, approximately 234 million operations are performed yearly. It's difficult to get a handle on the death-rate percentage from post-surgery complications, but I've seen estimates anywhere from 1.5 to 5 percent within the first thirty days after surgery. If we take the low end, that's more than 3.5 million post-surgery deaths a year. Of that number, a significant percentage—perhaps as high as 50 percent—is attributable to infections and complications that could have been prevented if safety procedures had been followed. This study included eight hospitals in eight cities, including Toronto, New Delhi, Manila, London, and Seattle—a socioeconomic and cultural cross-section of hospitals that participated in the World Health Organization's Safe Surgery Saves Lives program. As a benchmark, data was collected on 3,733 consecutively enrolled patients sixteen years of age or older who were undergoing noncardiac surgery. After the surgical-safety checklist was introduced, researchers collected data on 3,955 consecutively enrolled patients with the same criteria. Both the death rate and the overall complications rate were analyzed for the first thirty days after the operation. The results: The rate of death at the hospitals was 1.5 percent before the checklist was introduced and declined to 0.8 percent afterward. The rate of complications was 11 percent before the checklist, and declined to 7 percent after. In short, using the checklist cut the death and complications rates nearly in half.

This basic tool simply made people remember to follow the safety procedures every time, resulting in fewer complications and fewer patient deaths.

Two weeks later, participants were brought back and given information on four of the events they'd previously rated in the low-to-moderate plausibility range (no UFOs). The information included newspaper articles, third-person descriptions, and data from previous study subjects—all of which were designed to establish higher prevalence of the events. The results showed that high-prevalence information from all sources affected the development of false beliefs. In particular, participants given high-prevalence information in false newspaper articles became more confident that they had actually experienced the events, testifying to the power of the printed word on memory.

The takeaway here probably has a couple of prongs. First, we shouldn't discount the possibility that we're just as susceptible to developing false beliefs as anyone else walking around on this planet. The brain is a superb miracle of errors, and no one, except the brainless, is exempt. But knowing this about our brain is probably the best preventive against chasing the make-believe rabbit down its hole, if we can remember to keep this realization in mind.

LAST WORD: TOTAL FUTURE RECALL

If you stop and think about it, the ability to construct future scenarios in our minds is really quite remarkable. As far as we know, all other species, including our closest primate relatives, react to events as they occur. They can learn from these events and apply that learning in the future (think of chimpanzees learning how to catch ants by poking a stick into an ant mound and doing the same thing for every new mound they find), but they do not assemble complex chunks of information into even more complex and coherent swaths of the future. How exactly do we accomplish this?

As with many issues in cognitive science, it is difficult to say for certain, but the latest thinking is that we engage in something called *episodic future thinking*, which means that we simulate the future by using elements from the past. Recent brain imaging studies show that some of the brain regions that are activated when recalling a personal memory—

the posterior cingulate gyrus, parahippocampal gyrus, and left occipital lobe—are also active when thinking about a future event.[8]

Our future simulations are obviously not carbon-copy replicas of the past—but we draw on experience to generate the simulations in the same way that a sketch artist uses the pieces of information provided to her to compose an image. Sometimes we get close, other times we are far off; the farther removed the future scenario is from actual experience, the less likely it is to be in the proverbial ballpark.

Perhaps this ability, among other abstract-thinking abilities, confers an adaptive advantage. For the happy brain to make anywhere-near-accurate predictions about our environment, it helps to have access to as much past information as possible to construct multiple scenarios of what *could* happen. It could be that what we lack in instinctual gumption, like that of other primates, we make up for in imagination. The ability is far from perfect, but it is the most powerful organically based prediction tool yet evolved.

Chapter 14

BORN TO COPY,
LEARN TO PRACTICE

"Smooth ice is paradise for those who dance with expertise."
—FRIEDRICH NIETZSCHE, *THE GAY SCIENCE*

The Marx Brothers were masters of comic timing, as hilariously demonstrated in the 1933 movie *Duck Soup*. Groucho, Harpo, and Chico appear in the famous greasepaint eyebrows, mustache, and round glasses, while wearing nightcaps. They are so indistinguishable, it's almost impossible to tell them apart, which makes the famous "mirror scene" work perfectly. In the scene, Groucho stands on one side of a doorway and Harpo and Chico on the other, though not at the same time. Every move Groucho makes is imitated by one of the other two, creating the illusion that the doorway is actually a mirror. Groucho is suspicious and tries to throw off his reflection, but he's met move-for-move by the imitators for several minutes, until they finally make a mistake and both appear in the doorway at the same time.

I'm sure the Marx Brothers didn't realize that they were giving a comic illustration of a hardwired aspect of the human brain that is central to our growth and development. We are all born imitators, and a good chunk of our lives is spent playing out the evolutionary version of the mirror scene. A happy brain is happy to copy, and doing so is largely an automatic, as opposed to voluntary, response. Just to be clear, this is not the same as saying that we emerge from the womb like a blank sheet of paper ready to be scribbled on. But we are born ready to reproduce observed behaviors, and without this ability we would be truly lost in the woods.

Several other species aren't quite so lost when they enter the world. If you have ever had the opportunity to watch a horse give birth, it is remarkable to see the foal almost immediately try to stand just after emerging. The prewiring to get a fast start asserts itself with hardly a moment to spare. Baby chimpanzees are, like human babies, very vulnerable and unable to do much for themselves for about a year, but just after being born they are able to cling to their mother without having any example to follow. They are, in a sense, hardwired to cling, and the behavior automatically happens. We humans, on the other hand, are born with really only one noticeable capability: to observe. We cannot sit upright or cling or stand, but our powerful brains are doing something that, with time, proves significantly more beneficial than any of those abilities. We are the top-dog learners on the planet, and our cerebral reign starts early.

That's the good news. The not-so-good news is that our brains' fantastic ability to learn by imitation can become a handicap when we veer into the stratum of overimitation. Added to that, a happy brain does not seem to have an especially effective on/off switch when it comes to imitation. When someone completely lacks the ability to regulate imitation, they suffer from a disorder known as echopraxia; their brain simply does not have an inhibiting factor to prevent imitation of others' actions. This disorder is readily found in those suffering from autism. The distinction between the self and the other appears to be blurred in those with echopraxia, almost as if the brain cannot tell if it is looking at a reflection or another person. Most of us, of course, do not have echopraxia, but our normally functioning brains are marginally handicapped nonetheless.

WHAT KIDS TEACH US ABOUT IMITATION

Psychology research on imitation suggests that when children are instructed to repeat an adult's behavior, they aren't doing so only to reach the goal—the behavior itself becomes as important as the goal. This may be because human children can be easily persuaded to believe what adults tell them, even when it contradicts their own senses or because our most potent hardwired learning strategy is imitation—even when such imitation seems illogical.

A study titled "The Hidden Structure of Overimitation" in the *Proceedings of the National Academy of Sciences* focused on that last point. Using a "repeat-this-action" method, researchers tried to gain a better understanding as to why children will repeat irrelevant adult actions, something even chimps aren't prone to do. Kids in the study were told to imitate an adult's actions to retrieve an object from a "puzzle box" that had a variety of strange gizmos attached, none of which were needed to get the object. Even when given explicit directions to simply retrieve the object—and it was obvious that only one action was needed to do so—the kids would imitate every action taken by the adults, pointlessly pulling and pressing the gizmos before grabbing the prize just as the adults did. When the same test was conducted with chimps, they simply retrieved the object as instructed. Researchers concluded that this tendency is more than a human social dynamic in the making—it's a cognitive encoding process at the core of how we learn, and it comes at a cost. Children who observe an adult intentionally manipulating an object have a strong tendency to encode all of the adult's actions as meaningful. From the study: "This automatic causal encoding process allows children to rapidly calibrate their causal beliefs about even the most opaque physical systems, but it also carries a cost. When some of the adult's purposeful actions are unnecessary—even transparently so—children are highly prone to misencoding them as significant."[1]

So this study suggests that the same encoding process that allows us to develop a sense of an action's significance also makes us prone, as children, to misencoding purposeless actions as causally significant. This effect is so potent that, once engaged, it's extremely difficult to break.

What this means in the context of learning through practice is that humans begin life with a propensity to learn the wrong way (as well as the right way). Bad lessons are learned as readily as good ones, and we may not even know the difference.

The Brain That Mirrors

One of the more exciting brain science discoveries of the last few decades is that of a type of neuron that seems designed to "mirror" the actions of others. While most of the research-based evidence for their existence has so far come from studies with monkeys, more evidence is accruing for a possible "mirror system" in the brains of humans.

A typical example from a monkey study shows that the same neurons that fire when a monkey picks up a ball also fire when it watches someone else pick up a ball. The neurons appears to encode the action in such a way that they "reflect" the actions of others by responding the same way in the observer's brain. There are multiple theories as to why these neurons exist, but the general thought is that they facilitate learning by observation. In monkeys, this seems true with respect to mirroring actions, but the possibilities in humans are even broader. It's possible that the mirror system plays a role in all sorts of human learning, including understanding others' intentions, language acquisition, and even empathy.

Now we will shift the discussion to applications of learning through practice, starting with the now infamous specter of "expertise."

WHAT WE HEAR ABOUT EXPERTISE, AND WHY IT DOESN'T REALLY MATTER

Much has been said in recent years about expertise and the time and effort requirements of attaining it, and there are several books out there that dive into the details more than I plan to do so here. We have heard about the "ten-thousand-hour rule," that expertise requires at least ten thousand hours of practice.[2] I won't spend any time challenging or affirming the rule (its specific accuracy has been challenged in other sources); suffice it to say that whatever the actual amount of time required to gain expertise, it is a substantial, multiyear investment. Indeed, it may consume a massive portion of our lives.

Clearly, though, time is not the only requirement. Years of one's life spent practicing the wrong things will not lead to expertise any more than spending the same amount of time watching television. As a colleague of mine likes to say, "You can spend a lot of time getting good at being wrong." Time is a basic prerequisite, but not a sufficient one in itself. Layered upon time are a slew of other ingredients, like focus, precision, discipline, and desire—to say nothing of effective teachers along the way. It also doesn't hurt to have a few breaks go your way as well.

We will leave the discussion of "expertise" there, because in my view it is more a "topic de jour" than it is a topic worth talking about. More basic to most of our lives is how to gain a level of mastery in the things we do, or would like to do, without burning ourselves out or giving up prematurely. And what I want to argue is that those two pitfalls are always very real possibilities for the happy brain, chiefly because we have to push against some stubborn in-built tendencies to get where we want to go.

YOU DON'T KNOW WHAT YOU DON'T KNOW

Anyone who has ever been hired into a position he is not qualified for is in touch with a horribly awkward sensation, one akin to feeling lost. What we thought we knew about how to perform in the position turns out to be wrong or incomplete. The things we need to know in order to do the job are not clear. The phrase *over your head* is appropriate, because truly what we don't know is threatening to engulf us. People who open restaurants for the first time are known to say that they had no idea what they were getting into. From the outside looking in, they saw only one thin slice of an enormous undertaking. When they started coming face-to-face with those facets of the business they did not even know existed, the reality that they were barely toddlers in a game of grown-ups started settling in.

Of course, it is preferable to start somewhere rather than not start at all, and that's exactly what most of us do. But we are fooling ourselves to think we will saunter into a new role, job, trade, or any other position fully equipped with what we need to succeed.

What we should be thinking, rather, is how to ramp up as quickly as possible to the needs of the position. That only happens through what expertise scholar K. Anders Ericsson calls *deliberate practice*—practice designed to develop mastery in the specific areas required by whatever role or position we are targeting.[3] Whether one puts in ten thousand hours or fifty thousand hours, without deliberate practice the effort will still fall short. This is a crucial point, because energy wasted on poorly directed practice is a drain on the brain—and the more time you spend in that mode without results to show for it, the greater the tendency to burn out and give up.

TO GENERALIZE OR SPECIALIZE . . . *THAT* IS THE QUESTION

Central to the topic of deliberate practice (or what I will simply call *practicing for the purpose*) is the question of whether general or specialized problem-solving strategies are more effective. It is an important question with implications for how skills are taught—most important, thinking. General problem-solving strategies are context-independent. For example, if Charley the policeman is taught the general strategy for safely disarming criminals, then he should be able to apply it to a general range of situations in which he faces armed criminals. The general strategy, this argument goes, produces a generally applicable problem-solving ability.

On the other hand, with his general strategy for disarming criminals, Charley may in fact be fairly effective—unless and until he encounters a specific situation that trumps his general ability. He may, for example, be expert at taking away a criminal's handgun—but what happens if he encounters a criminal with another gun tucked in the back of his pants and a knife concealed in his sock? If Charley never faced or trained for *that* situation (or one very much like it), he may be in for more than his problem-solving strategy can handle.

How Magicians Practice Imitation to Create Illusion

Magicians fool us with sleight of hand—the ability to make it seem as if they are reaching for or placing an object that isn't really there. Most of us can't pull this off, which makes the magicians' craft all the more astounding. Research published in the open-access journal *PLoS ONE* used a novel approach—motion-tracking technology—to reveal how magicians deceive our eyes.[4]

Ten nonmagicians and ten magicians were presented with a wooden block on a table and asked to reach for and pick it up, or to pretend to pick up an imaginary block next to the real one. As the participants reached for the block, their sight was obscured (to simulate a magician's technique of turning away from the object). Motion tracking showed that nonmagicians' grasping technique changed from the real block to the imaginary one, but the magicians' grasp was identical for each.

In a follow-up experiment, participants were asked to grasp for a battery, but during the imaginary-object session, the real object was removed from the table. Under these conditions the magicians failed just as often as nonmagicians. These experiments suggest that magicians use visual input from real objects to calibrate their grasp of imaginary ones. With time and practice, magicians develop flexibility in a part of the brain associated with spatial reasoning and learn to precisely adapt their movements so seamlessly that our eyes fail to catch them in the act.

A study published in the journal *Cognitive Science* backs up the argument for specialization, albeit with a less exciting example than disarming criminals. Expert chess players, specialized in different openings, were asked to recall positions and solve problems within and outside their area of specialization.[5] All of the players' general expertise was at roughly the same level. The results: Players performed significantly better in their area of specialization, but not only better—they actually played over their own heads at the level of chess players with much better general skills. In other words, specialization trumped general problem solving *and* elevated the players' level of play.

This study suggests that when figuring out how to tackle a problem, knowledge derived from familiarity with *that* problem is more important than general problem-solving strategies. The key is memory. We rely on memory of specific experiences to craft solutions to new problems. If you have expert general ability but lack context-specific memory, you're only as effective as general ability will allow—and if you're Charley, that might not be good enough.

PRACTICE LESSONS FROM TAXI DRIVERS AND BURGLARS

London taxi drivers have to undertake years of intense training known as "the knowledge" to gain their operating license, including learning the layout of over twenty-five thousand of the city's streets. A study by psychologist Katherine Woollett and her team investigated whether their expertise can be effectively generalized in new situations.[6]

The taxi drivers, and a control group, were first asked to watch videos of routes unfamiliar to them through a town in Ireland. They were then asked to take a test about the video that included sketching out routes, identifying landmarks, and estimating distances between places. The taxi drivers and the control subjects both did well on much of the test, but the taxi drivers did significantly better on identifying new routes. This result suggests that the taxi drivers' mastery can be generalized to new and unknown areas. Their years of training and learning through deliberate practice prepare them to take on similar challenges even in places they do not know well or at all.

Burglars, as it turns out, have much in common with taxi drivers, at least in the sense of developing mastery. Researchers interviewed fifty jailed burglars, all of whom had committed at least twenty burglaries in the last three years; half had committed more than one hundred burglaries. The researchers asked questions regarding what technique the burglars used to search inside houses. Over three-quarters of them described searching as relatively routine, and many of them used terms such as *automatic* and *instinctive*. The study also found that on average, it took an experienced burglar less than twenty minutes to identify everything in the house worth stealing and walk out with it. For the most part, the burglars said, they didn't really have to think about what they were doing. Years of practice on similar houses with similar layouts and possessions enabled autopilot to take over while the burglars went to work. While not evidenced by terrific role models by any means, from a mastery standpoint these results are (uncomfortably) enlightening.

PARTING THOUGHTS

Coaches like to tell their players to "work smart," and that is actually a fitting phrase to end this chapter. The happy brain is willing to work, but when we plow into a new endeavor without a sense of direction and purpose in our practice, negative results are sure to come: namely burnout, disillusionment, and eventually giving up. Better to instill focus early on and seek out whatever assistance we need to make our practice purposeful and worthwhile.

Part 6

NOTHING SO PURE AS ACTION

Chapter 15

MIND THE GAP

"Action is character."

—F. SCOTT FITZGERALD,
FROM NOTES TO AN UNFINISHED NOVEL,
THE LOVE OF THE LAST TYCOON

I'm standing near the deli counter at the supermarket. Close to me are five or six other people, and we are all eyeing the same quarry—rotisserie chickens turning on a spit in the monster-sized oven against the far wall. The timer on the oven tells us that there are just over three minutes left before the chickens are ready. More people gather. I inch closer to the counter. The others do the same. I can feel a tension thickening in the atmosphere; my nerves are starting to peak, my heart is beating faster. I stop for a second and ask myself why I feel this way. All of us standing at the counter can see that there are more than enough chickens available for the crowd. Even if there weren't, we are in a grocery store full of food. No one is going to starve in this scenario. In addition, we are all adults capable of civilly dividing the chickens among us when they arrive. None of us will have to fight for our food or risk our lives against other predators to secure our families' sustenance. And yet, the tension persists.

What I know for sure is that I cannot undo my brain's tendency to amp my energy level and alertness to ensure that I get the food I came for—a tendency that has been neurally hardwired, for good reason, and been shown to be valuable under certain conditions. What I can do is recognize what is happening, identify why the reaction is out of place for the circumstances, and relax.

That simple example illustrates the two things that must happen for us to effectively address the problematic tendencies of a happy brain: *elevate awareness* and *take action*. Awareness of why we are doing what we are doing is a crucial step toward action because it initiates a change in thinking—we have to pause to examine what's going on. And this is why *science*-help is more useful than typical *self*-help. Gleaning evidence-based clues from cognitive science provides tools to bolster awareness and enable action. I'm certainly not arguing that they provide a foolproof road map for action, but my hope is that after reading a book like this one you will have more usable knowledge to work with than you did before. The vital point to remember is that the gap between knowing and doing is ever present until we commit to acting.

For the rest of this chapter I am going to offer a selection of knowledge clues drawn from research discussed throughout the book. Many of them build on discussions from earlier chapters; a few are new to this one.

SLOW DOWN.

So many problems can be diffused by slowing down and carefully considering how to proceed in any given situation. In some instances, of course, there isn't time to slow down and we have to just react. But generally we have more time than we allot ourselves to make decisions and draw conclusions. Putting on the mental brakes can stop you, for example, from reacting in anger to someone on the road—a situation that can rapidly escalate out of control. Slowing down provides time to consider how an issue has been framed and whether we have really considered all of the relevant factors. Pausing for just a moment can allow you to challenge yourself about an action you are about to take that could have horrible consequences, like responding to an email on your smartphone while driving instead of waiting until you can focus attention on the message you want to send—instead of parsing attention between the email and driving. Slowing down is, in short, fundamental to nearly every topic in this book. If more of us would take just a couple extra moments to think an action through, we would all be much better off.

BE AWARE OF THE INFLUENCE YOUR PREEXISTING BELIEFS ARE EXERTING ON YOUR CURRENT THINKING.

We are all biased thinkers. No one is a "blank slate," and therefore no one's perception is free from the influence of preexisting beliefs. The question is, are we aware that this is the case? Racists often justify their comments by saying, "That's how I was raised." And that may be true. Childhood imprinting is a major source of preexisting beliefs, and most are resistant to deconstruction. But forcing ourselves to be aware of this influence on a case-by-case basis can, with time, challenge the entire infrastructure of questionable belief. Cognitive-psychology research reinforces this message again and again by showing that incremental change is a more effective way to go than attempting exhaustive change. Challenging ingrained patterns of thought takes work, time, and persistence.

CHECK YOUR AVAILABILITY BIAS.

As mentioned earlier in the book, a happy brain tends to make judgments using the most accessible and available information. For example, people typically judge the incidence of crime as much higher than it actually is. The reason cited by psychology researchers is that the news media focuses on crime, thus increasing its availability and accessibility to the audience. The same thing happens when you expose yourself to one perspective and ignore others; the availability of that perspective (about politics, for instance) leads to a bias that the chosen perspective is the correct one. If all of the radio talk shows you listen to trumpet generally one perspective, we can safely predict how you will respond to different positions. Simply knowing this can be enough to challenge thinking, but frequently this bias is connected to related tendencies like confirmation bias and framing (discussed in chapter 1), and substantial effort, along with more than a little humility, is needed to move in another direction.

BECOME SAVVY ABOUT FRAMING.

Earlier we spoke of the brain's tendency to sound alarms when focus moves outside the parameters of a perceptual "frame." Perhaps while you were growing up, your parents and siblings told you and everyone else that you were the "smart one" in the family, while your brother was the "athletic one." You didn't realize that years of ingesting these labels framed your self-perception. Without ever really challenging the point, you simply expected to be less athletic than your brother, and that influenced you to not participate in sports and to instead spend time being the smart kid who gets all As. Said another way, you were complicit in keeping in place the "not athletic" frame created by your family. Thinking differently of yourself just doesn't feel right; in fact, it makes you anxious and uncomfortable. Breaking a deeply internalized frame like that is extremely difficult—just becoming aware that it exists is a major step. On a more day-to-day basis, we encounter dubious framing all of the time; the way statistics are presented, or the way an argument is structured, for example. Developing the skill to deconstruct the frame allows us to see alternative explanations, and can also keep us from getting scammed by the unscrupulous.

ENGAGE OTHERS TO HELP KEEP YOU ACCOUNTABLE TO YOUR COMMITMENTS.

It is important not to confuse this statement with asking others to *help you keep* your commitments. The suggestion here is that when you make your commitments "public" with a select few friends and/or family, and ask them to check in with you on progress, you are giving yourself even more incentive to reach your goals, assuming the opinions of those people who matter to you. This suggestion builds from the realization that we are an interdependent species, not a wholly independent one. For most of us, asking others whom we respect to inject some accountability into our commitments can help produce better results.

ACT ON SHORT-TERM REWARDS THAT WILL EVENTUALLY YIELD LONG-TERM BENEFITS.

As we have discussed, the happy brain tends to focus on the short term. That being the case, it's a good idea to consider what short-term goals we can accomplish that will eventually lead to accomplishing long-term goals. For instance, if you want to lose thirty pounds in six months (just in time for swimsuit season), what short-term goals can you associate with losing the smaller increments of weight that will get you there? Maybe it's something as simple as rewarding yourself each week that you lose two pounds. The same thinking can be applied to any number of goals, like quitting smoking or improving performance at work. By breaking the overall goal into smaller, shorter-term parts, we can focus on incremental accomplishments instead of being overwhelmed by the enormity of the goal.

MAKE GOALS TANGIBLE AND MEASURABLE.

Continuing the discussion of goals, it helps to make them tangible (recalling that the happy brain is value- and reward-oriented). Losing weight has an obvious tangible and measurable result. Benefits from quitting smoking may not be as obvious right away, but tracking how you felt before you quit and how you feel a few months later makes the goal tangible. This little trick is called *feedback analysis*. Simply taking a few notes on yourself (on your physical and mental well-being, for example, not to mention how much money you have saved) and then revisiting those notes a few weeks or months later will provide tangible evidence of the goal's value. The ways in which this tool can be applied are limitless, provided that your pre- and post-goal assessments are honest and, to the best of your ability, unbiased.

REMEMBER, THE HUNT IS MORE EXCITING THAN THE CAPTURE.

One of our brains' especially frustrating habits is to focus on getting a reward, and then experience a feeling of loss once we get it. This cycle can spin us into a loop of wanting, getting, and regretting. Awareness that you are caught in the cycle is paramount. If you are bidding on items online and find yourself compelled to keep bidding up the price of an item beyond its value or what you intended to spend, force yourself to become aware that what you are doing is no longer in your best interest. The action part is harder, because you have to walk away from the target. If you don't, you cannot expect merely thinking rationally to correct the problem, because it rarely ever does. We are master justifiers, and almost any rational reason given (by ourselves or others) for stopping an action can be dismantled in minutes or less. Action in this case is necessarily absolute: stop and walk away. If the circumstances call for it, run.

ENVISION COMPETING FUTURE NARRATIVES, BUT CHECK YOUR SELF-SERVING BIAS.

We have seen that our brains have difficulty placing us in the future, which renders making sound decisions that impact the long term a hard thing to do. We have also discussed that memory has both "looking back" and "looking forward" dimensions. We tend to simulate the future by reconstructing the past, and the reconstruction is rarely accurate. What can we do? Envision competing narratives of the future—good and bad. At the same time, make sure that we aren't coloring those narratives with what psychologists call *self-serving bias*, which leads us to believe that our success will be due to our actions, and any failures will result from other factors.[1] For example, a student studies for an exam and is convinced that if he earns a high grade it will be due to his intelligence and hard work. If he doesn't get a high grade, it will only be because the exam was poorly written or the instructor weighted the grades unfairly. The student's future narrative has only one dimension: Hard work and intelligence *will* result in a high grade. While it may seem odd to suggest envisioning failure at our own hands, it's not a bad grounding exercise to avoid this sort of one-dimensional thinking. Failure is a possibility and should be recognized as such, even though we are obviously designing our actions for success.

PRACTICE FOR THE PURPOSE.

In business circles, the term *strategic thinking* is thrown around with abandon. It's a very general term that really only takes on meaning when applied to a specific situation. Strategic thinking in a marketing context is not the same as strategic thinking in a financing context. As a general proposition, the term is vague at best. As a specific application, it can be indispensable. Much the same is true of practice. Practicing to become faster has benefits, no doubt, but those benefits may not apply to every situation. Being fast in football is not necessarily the same as being fast at a track meet. In football, the application of "fast" likely includes knowing how to make sharp cuts while running, following routes, or avoiding being tackled. For a track runner, those skills are meaningless. Practicing for the purpose takes the best of generally applicable skills and combines them with specific applications. Without the application, your preparation isn't even halfway complete. The caveat to remember is that it is also possible to overspecialize, in which case your brain will strain to apply your learning to new situations.

FINISH WHAT YOU START.

This suggestion is one in which leveraging a tendency of a happy brain works in our favor. Incompleteness represents instability to our brain. A basic illustration of this is the open-circle experiment. Draw a circle on a piece of paper, but leave a small gap such that the circle is not all the way closed. Now stare at it for a couple of minutes and notice what happens—your brain wants to close the circle. For some people the urge to close it is so strong that they'll eventually pick up the pencil and draw it closed. The same dynamic can be applied to stop procrastination from burning you. The trick is simply to start whatever project is front of you—just start anywhere. Psychologists call this the *Zeigarnik effect*, named for the Russian psychologist who first documented the finding that when someone is faced with an overwhelming goal and is procrastinating as a result, getting started anywhere will launch motivation to finish what was started.[2] When you start a project—even if you begin with the smallest, simplest part—you begin drawing the circle. Then move on to another part (draw more of the circle), and another (more circle), and so forth. The one prerequisite for the Zeigarnik effect to work is that you must be motivated to complete the project in the first place.

ASK, DON'T TELL (YOURSELF).

Self-motivation isn't easy, but a few subtle tweaks can make it easier and more productive. Psychology research suggests that *asking* yourself if you can accomplish a goal is more effective than *telling* yourself you will. You'll recall the comparison between the Little Engine That Could telling himself, "I think I can, I think I can," and Bob the Builder, whose mantra is "Can we fix it? Yes we can!" As it turns out, Bob's approach is more productive from a motivational standpoint.

FORM USEFUL HABITS.

Psychology research tells us that the average amount of time necessary to reach *maximum automaticity* (a jargony term for *habit*) is sixty-six days. But when you are trying to develop a healthy habit, it's likely that it will take at least eighty days for it to become automatic. The more complex the habit, the longer it takes to form. An exercise regimen, for example, will take most people at least one and a half times longer than more basic endeavors like changing eating or drinking habits (which still take a long time). You can afford to miss a day here or there, but the more cumulative days you miss, the more habit formation is disrupted. The crucial point to remember is that creating useful habits is as important as discontinuing bad ones—and worth the effort.

HOW YOU WANT OTHERS TO SEE YOU CHANGES YOUR FIRST IMPRESSION.

We discussed in chapter 9 that our brains interpret first impressions as value propositions. In addition, we are prone to judging the first impression someone gives us by the standard we've set for ourselves. So if we want to come across as gregarious and fun loving, we are also evaluating others by that measure. If they fall short, it may be because they aren't measuring up to an artificially high standard. But what if we don't have a particular impression in mind when meeting someone new? The research suggests that we are still making a value assessment, and in part this assessment is about trust. From the get-go, our brain is making calculations as to whether the person in front of us can deliver on a trust exchange or if something isn't quite adding up.

TRY TO REMEMBER, YOUR MEMORY JUST MIGHT BE WRONG.

Memory is not a recording, it's a reconstruction. We are prone to "confabulate" pieces of actual memory with other information, and a happy brain assembles the parts into something we can easily mistake for a flawless recollection. Most of us will not realize that our memory of an event lacks information that has been supplemented by our brains from other sources. This has huge implications for eyewitness testimony—and for you when you are arguing with your spouse about something you claim to remember perfectly. Better to swallow the fact now that "perfect memory" shares the same handicap of everything that is "perfect"—it isn't.

HABITUATION HAPPENS.

It's a cruel fact of human existence that with enough time, we can become bored with just about anything. Whether it's a new car or a new dog, a great Indian dish or a great song, eventually the initial pleasure fades into something more mundane—which doesn't necessarily mean we come to dislike the thing in question, but rather that we habituate to its once-tantalizing allure and simply enjoy it less. Even sex (gasp!) isn't immune.

MIND THE GAP

The challenge we face is overcoming "variety amnesia"—our tendency to forget that we've been exposed to a variety of great things, be they people, food, music, movies, home furnishings, and so on—and instead focus our attention on the singular thing that no longer gives us the tingles. To shake ourselves free from this negative trap, we must "dishabituate" by forcing ourselves to remember the variety of things we've experienced. For example, let's say that you've become bored with a particular musical group you once couldn't listen to enough. Research suggests that what you need to do is recall the variety of other songs from other musical groups that you've listened to since the last time you listened to your once-favorite band, and by doing so you'll revive appreciation for your fave. Psychology researchers call this little head trick a simulation of *virtual variety*, which reduces satiation—the lessening of satisfaction over time—in a way similar to that of experiencing actual variety.

IMAGINE EATING THE TREAT TO SHORT-CIRCUIT FOOD TEMPTATIONS.

If you imagine looking at a tempting treat, your desire for it will increase. But research indicates that if you imagine eating the same treat, your desire will lessen. The reason is that to our brains, imagining an action and doing it are not too dissimilar. We can trick ourselves into feeling like we've already enjoyed the treat, leaving our brain with less reason to target the genuine article.

EMPATHY ISN'T BLIND.

Remember the yawning chimps from chapter 11? The gist of that research is simply that we empathize much more with those familiar to us, and this familiarity bias is demonstrated in something as basic as contagious yawning. It's also evidenced by contagious laughing and crying, among other behaviors. Our ability to empathize with others is an important one, but if you notice, it's far easier to empathize with someone you know than with a stranger. The reason is that our brains evolved to socially relate to a relatively small group. Some specialists in this area put that number at roughly 150 members. Beyond that, we draw more distinct lines and are selective about investing emotional energy. Unfortunately, this same tendency can make us callous to the plight of those who need help in other parts of our country and around the world.

PRACTICE METACOGNITION.

Metacognition simply means "thinking about thinking," and it is one of the main distinctions between the human brain and that of other species. Our ability to stand high on a ladder above our normal thinking processes and evaluate why we are thinking as we are thinking is an evolutionary marvel. We have this ability because the most recently developed part of the human brain—the prefrontal cortex—enables self-reflective, abstract thought. We can think about ourselves as if we are not part of ourselves. Research on primate behavior indicates that even our closest cousins, the chimpanzees, lack this ability (although they possess some self-reflective abilities, like being able to identify themselves in a mirror instead of thinking the reflection is another chimp). The ability is a double-edged sword, because while it allows us to evaluate why we are thinking what we are thinking, it also puts us in touch with difficult existential questions that can easily become obsessions. For the purpose of this book, it is important to recognize metacognition as an essential part of nearly everything we have discussed.

DON'T MAKE IMPORTANT DECISIONS
WHEN YOUR BRAIN IS ON EMPTY.

It's always useful to keep in mind that our brains use a great deal of energy to accomplish the impressive things they do. When we're running low on cerebral fuel, it is not the time to take on momentous challenges, if we have a choice. As a wealth of research illustrates, it seems that our brains hit an energy-drain threshold beyond which impulsiveness increases and sound decision making suffers. One study illustrated this by putting a group of volunteers through six hours of challenging memory tasks.[3] At various points while doing the tasks, the volunteers were faced with choosing either a small amount of cash now or a larger amount later. As the hours of memory challenges ticked by, the volunteers were increasingly more susceptible to taking the easy money. The researchers compared that group with a group of volunteers who did easy memory tasks (or spent the time reading and playing video games), and for the most part they didn't break down and choose the smaller amounts of cash up front. The researchers also scanned the volunteers' brains via fMRI and found that people doing the hardest memory tasks showed decreased activity in a brain area called the middle frontal gyrus, which other studies have shown is closely involved in decision making. It seems the dynamic at play is a form of brain fatigue; brain areas crucial to decision making run low on energy and succumb to fatigue, not unlike muscles after a difficult workout. We're all well advised to keep the muscle metaphor in mind when thinking about the brain, and not fooling ourselves into believing that it's somehow immune from the same fatigue that effects every part of us.

NOT ALL CONTAGIONS ARE PHYSICAL.

In our discussion of psychosocial contagions in chapter 9, we reviewed a handful of well-researched contagions, such as anxiety, blame, apathy, and happiness. Clearly there are good and bad aspects of catching or transmitting psychosocial contagions (most people I know wouldn't mind catching a little more happiness). The important point to remember is that awareness of the influence can prevent negative outcomes, particularly when fear, anxiety, blame, and anger are concerned. In a group setting, it is all too easy to be pulled into the viral spread of emotions that can, if left unchecked, lead to catastrophe. For example, consider retail store stampedes during the holidays, which often result in people being trampled to death. Or mob scenes at sporting events and concerts. These are all examples of psychosocial contagions gone awry, and anything we can do to halt their spread is worthwhile.

FEELING RIGHT IS NOT THE SAME AS BEING RIGHT.

One of the foibles we discussed in chapter 1 has to do with how certainty feels. A happy brain interprets uncertainty as a threat and wants us to get back to "right." But what we often overlook is that what we are really trying to recover is the *feeling* of being right—because it is the emotional response to rightness that shuts off the alarms and puts us at ease. It's easy to confuse this feeling with the real thing, and all of us are culpable. The truth, however, is that the evidence may not align with the source of your certainty, and that's a difficult realization for any of us to acknowledge.

KNOW WHEN TO ENGAGE HEURISTIC OVERRIDE.

Heuristics are simple, efficient rules—either hardwired in our brains or learned—that kick in on occasion, especially when we're facing problems with incomplete information. Happy brains like heuristics because they stifle uncertainty. They can be advantageous tools, but they can also be misleading handicaps. For example, when we are trying to decide which route to take on a trip and have several options to pick from, it is beneficial to rely on a simple heuristic guideline that says, "The most direct route with the shortest distance is usually best." But when we are struggling to determine whether to tell a friend that he is acting recklessly—and know we are risking the friendship—or to just try to support him through a rough time and possibly preserve the friendship, there is no basic heuristic that will illuminate the way. The decision is case-specific, and the choices both have pros and cons that must be weighed against each other; and even then the decision will be difficult. Our challenge is to know when heuristics are useful, when they are not, and when they will actually make a situation worse. If we can't use them to aid in the decision, it's time to engage heuristic override and move on to the tough work of struggling with ambiguity.

WE ARE COMPELLED TO CONNECT.

The human brain evolved to identify patterns by connecting pieces of information in our environment. While this hardwired habit has been vital to our survival and generally serves us well, it also leads us to make something out of nothing. We routinely make connections between chance events and assign meaning to randomness. These are not tendencies we can simply discontinue, but becoming aware of our compulsion to connect puts us in a better position to check ourselves when our pattern-finding penchant goes over the top. This is no trivial issue. People spend enormous time and money investing themselves in complex belief systems built on little more than coincidental toothpicks. The key is to value our brain's remarkable capacity for pattern detection while exercising vigilance about how we apply this ability in our lives.

WE CRAVE AGENCY.

Seeking "agency," as we discussed in chapter 2, refers to our desire to identify a responsible party for the good and bad events we and others experience. If something horrible happens and there is no apparent agent behind it, our brain will search one out regardless. Statements like "Everything happens for a reason" imply agency, because presumably someone or something has preordained the "reason." More direct positions such as "This is part of God's plan" tag a divine, personal agent as the cause—even if the rationale for why that agent would want such a thing to happen is inexplicable. One of the hardest things for us to accept is that many things happen without any form of agency. The thought of this alone is enough to put the happy brain on guard, because it opens the door to uncertainty.

DOING ANYTHING AT ALL IS A GOAL IN ITSELF.

We tend to think of goals as being specific, but research suggests that a broader goal of simply staying busy also inspires us. What this boils down to is that doing something is better than doing nothing, all else being equal. The "something" can be anything, no matter how trivial it seems to others. This finding echoes a quote often attributed to Mahatma Gandhi: "Whatever you do will be insignificant, but it is very important that you do it."

WE ARE NOT VERY GOOD EMOTIONAL FORECASTERS.

When we try to place ourselves in a future situation that is more emotionally charged than the one we are in now, we fall prey to *intensity bias*—a skewing of perspective that leads us to believe we can forecast how we would react under emotionally charged conditions. This is why it's very easy to say, "If I had been in that situation, I would have . . ." But unless we have been in the same or a similar situation before, we really have no credible basis for knowing how we would react.

FASTER FEEDBACK IS JET FUEL FOR PERFORMANCE.

Fear of disappointment is a powerful motivator. Combined with the fact that we are less prone to worry about outcomes in the distance, the timing of feedback becomes an important element in the motivation mix. When feedback is out in the distance, it feels less relevant, and therefore less deserving of a happy brain's attention. But when feedback is immediate, we focus our attention on what's coming around the corner and consequently put more energy into our performance.

OUR LIE-DETECTION SKILLS ARE ON PAR WITH ROLLING DICE.

Dr. Paul Eckman has dedicated his career to understanding the verbal and nonverbal clues people give when lying and how effective we are in identifying liars.[4] His conclusion is that most of us may as well just guess if someone is lying, because even when we try in earnest to identify liars, we do no better than chance. It's alarming to realize just how bad we are at catching liars, and it's equally alarming to know that trying to be empathetic can make us even worse lie detectors. Research suggests that mimicking another's behavior, one component of empathy, makes you even more susceptible to deception than you were already. This doesn't mean that we should not be empathetic, but rather that we should be careful about being too empathetic too quickly.

MAKE CHECKLISTS AND USE THEM.

Since memory is, as we discussed, more fallible than most of us realize, it helps to have memory props on hand to help out. Checklists are a simple but effective tool to keep holes in memory from wrecking your day or your life. You can read any number of books about using lists as part of a time-management system, but my point, in the book you are reading right now, is simply that, no matter what system you use, the basic purpose of the list remains the same: to keep you from falling prey to the imperfection of memory. Make checklists and use them.

COUNTERFACTUAL THINKING IS A DANGEROUSLY VALUABLE SKILL.

All of us have a tendency to look back on a decision and think, "If only I had chosen differently, I'd be better off now." We come to this determination by imagining that the facts of the decision *could have been* different than they really were. For example, you might think, "If I had taken the job in New York five years ago, then I would have been able to network more effectively and my career would be more fulfilling now." Perhaps that statement is 100 percent accurate. But more likely it's partly accurate and partly a fabrication. It lacks consideration of all of the variables you had to consider at the time and the fact that you spent days digesting those variables to come up with an answer. The fact is, with the information you had, you decided not to take the job in New York, and the decision can't be reversed. We think counterfactually for a very good reason; namely, it allows us to learn from our missteps and make better choices in the future. This is a deeply embedded survival skill of the happy brain, and we should feel fortunate that it is. But when this skill is misapplied and we start dwelling on counterfactual comparisons, the result is not going to be pleasant (ranging from generally dark emotions to serious depression). Our challenge is to know when to throw the red flag and stop ourselves before the dwelling begins.

REPETITION IS THE MOTHER OF PERSUASION.

Every election year, I dread the onslaught of political advertisements. I try my best to avoid them and can't wait until they end. But the one thing I do pay attention to is whether or not the messages repeated the most become major factors in the election. They almost always do. Does it matter if what one candidate is saying about the other is factual? Sometimes, but generally it does not. What matters most is that the message has been repeated enough to color public perception. And the really interesting part is that the public doesn't even have to be paying much attention—in fact, the less attention we invest in deciphering a message, the more likely we are to believe it. The more we focus on the message and deconstruct it, the more likely we are to be skeptical. More of the latter approach is desperately needed.

METAPHOR IS POWERFUL MEDICINE.

Seasoned speechwriters know that the best way to convey a message is with metaphors the audience can grab and understand with little effort. Metaphors can make multifaceted concepts seem simple and vague ideas seem relevant. With very few words, they can change the way we think about difficult topics. Research suggests that when metaphors are used to frame a discussion (like whether crime should be viewed as a "beast" or a "virus," referring back to chapter 10), the rest of the discussion will be viewed through the lens of the metaphor. Being mindful of how metaphors are used in what we read and hear puts us in a better position to evaluate what's *really* being said—or not being said.

YOUR BRAIN IS NOT ONLY IN YOUR HEAD.

The theory of "embodied cognition," that our bodies are active participants in cognition, has been steadily gaining momentum. Research continues to show that the mind is influenced by the weight, size, texture, taste, temperature, and other attributes of physical objects. Experiencing weight in a physical sense, for example, influences our perception of "weightiness" in a perceptual sense. What this means is that influences on our thinking surround us all of the time and we don't realize it, though exactly why this happens is still not entirely clear. What is clear is that the mind—what our nervous system *does*—is never fully isolated from the world around us. Rather, it is constantly interacting with the environment, and this interaction is integral to how we think.

YOU DON'T KNOW WHAT YOU DON'T KNOW.

Obvious as this may sound, in practice it's not so obvious at all. It's tempting to think that we can jump right into an occupation and perform just as competently as anyone else. Perhaps the others have been practicing their trade for years. But still we think, "Why waste all that time when I can just start doing it?" The reason underlying this thinking is that we don't know what we don't know. Practice and experience aren't just preparation—they are part of a process of discovering what you could not possibly know as an outsider to whatever trade or profession you hope to become competent in. The phrase *taking your knocks* is a rough but accurate way to describe the process.

COGNITIVE FLUENCY ENABLES LEARNING, BUT ALSO PROPAGANDIZING.

When information is presented to us in an accessible, easy-to-digest way, our brain experiences fewer obstacles to processing it—even if the content is complex. Another way to say the same thing is that the more cognitive fluency information has, the easier it will find a home in our mind. Cognitive fluency is crucial to learning at all stages of life. But the same quality that makes it so important for learning also makes it a potent tool for persuaders of every sort, from advertisers to propagandists. The trick is to package the information in such a way that it links up with existing knowledge structures in your brain (aka *schemata*, referring back to chapter 1).

MORAL SELF-REGULATION IS A SEESAW PERFORMANCE.

Life is a balancing act, and so is our sense of morality. Research suggests that when we view ourselves as morally deficient in one part of our lives, we search for moral actions that will balance out the scale. Maybe you know you should be recycling but just never get around to gathering up your glass, paper, and plastics in time for the recycling truck. One day you happen to be walking through a hardware store and notice a rack of energy-efficient light bulbs, and you instantly decide to buy twenty of them and change out every bulb in your house. The moral deficiency (not recycling) is, in your view, balanced by a moral action (installing energy-efficient bulbs). The problem is that the seesaw can also tip the other way: If we believe we are doing enough, morally speaking, then there's little reason to do more. The scale is already level.

TO YOUR BRAIN, BELIEF JUDGMENTS LOOK THE SAME.

The brain evolved, as we've noted a few times throughout the book, to make sense of our environment and give us a better chance of surviving. Many of the applications of these broad adaptive skills seem to us very specific (the survival example in the supermarket deli that I started this chapter with is a basic illustration). But research indicates that the brain does not distinguish between many of the things that seem obviously distinct. The brain's reward system does not distinguish between what we in moral parlance would call "good" and "bad" rewards—it responds to rewards the same way despite our moral positions. This is also true of belief judgments; whether belief in the results of basic arithmetic or belief in God, the brain engages the judgment the same way.[5] Our day-to-day experience of the belief, of course, is much different depending on the subject—but that is a function of the meaning we assign to beliefs, not of the beliefs themselves.

MAKE PEACE WITH PROBABILITY.

Whether we think someone has "good luck" or "bad luck," in the end all so-called luck comes down to probability. It's tempting to interpret the outcomes of probability in such a way that it seems something was "meant to happen," but the truth is that winning the lottery or taking a direct hit from a hurricane are statistically explainable events regardless of how pleasant or horrific they are to experience. This is tough to accept, particularly for the human brain that craves certainty. Knowing that probability underlies everything we do does not necessarily make the outcomes any easier to swallow, but there is satisfaction in accepting the truth as it is without a veneer of mystification.

BE OF SERVICE TO OTHERS, EVEN IN SMALL WAYS.

If you want to reduce the effects of daily stress on your emotions and outlook, be of service to others any way you can. This has been shown in research that tested how small, helpful gestures throughout the day provide a significant boost against the negative impacts of stress.[6] The study included adults, ages eighteen to forty-four, who participated in a daily assessment via their smartphones for two weeks. Once a day, they reported any stressful events they experienced at work or home, or related to relationships, finances, or health. They also reported all of the helpful behaviors they engaged in during the day (even small gestures like holding open a door), and they completed a ten-item Positive and Negative Affect Scale, a well-validated measure of experienced emotion. They were also asked to rate their perceived mental health for each day using a sliding scale that ranged from 0 (poor) to 100 (excellent). The results were significant: the more helpful behaviors they performed each day, the higher their level of daily positive emotion and the better their overall mental health. Participants who reported lower levels of helping behaviors showed less resilience to daily stressors, evidenced by lower levels of positive emotion and lower self-evaluations of mental health. Helping others not only improves your mood but also provides a layer of emotional protection.

THINK TWICE BEFORE ACCEPTING NOMINAL VALUE.

Our brain is easily tricked into focusing on nominal value (or what we could call face value), and ignoring actual value. Nobel Award–winning psychologist Daniel Kahneman calls this susceptibility to focus on face value the *money illusion*.[7] If, for example, you are given a 2 percent raise in salary, but the rate of inflation has increased by 4 percent, you are actually in the hole by 2 percent. But that isn't how most of us see it. Instead, we focus on the dollar equivalent of a 2 percent raise, not the fact that the dollar amount is significantly less than the increase in cost of living. While it may be difficult, or impossible, to change the amount of your raise, it's a good idea to keep the face value versus real value lesson in mind whenever you are evaluating value, particularly when someone is trying to convince you that the face value tells the whole story.

DOUBT YOUR THOMAS.

The Human Oxytocin Mediated Attachment System (THOMAS) is a powerful brain circuit that releases the neurochemical oxytocin when we are trusted, and induces a desire to reciprocate the trust we have been shown—even with strangers.[8] Having a healthy THOMAS is good, because without it we would find it difficult to extend trust to others. But it's also a handicap, because the same system that enables trust makes us marks for con artists and criminals. Research indicates that about 2 percent of people are unconditional nonreciprocators: When trusted, they do not reciprocate with trustworthy actions (e.g., you trust someone enough to lend her money and she never returns it). What this means is that you will encounter people in your life who are good at engendering trust for the purpose of taking advantage of you. Your THOMAS isn't always wrong, of course, but it is wise to exercise vigilance just in case.

YOU MIGHT LOSE YOUR COOL, BUT DON'T LOSE PERSPECTIVE.

Loss aversion—or, simply, fear of loss—is a basic part of being human. To the brain, loss is a threat, and we naturally take measures to avoid it. We cannot, however, avoid it indefinitely. One way to face loss is with the perspective of a stock trader. Traders accept the possibility of loss as part of the game, not the end of the game. What guides this thinking is a "portfolio" approach; wins and losses will both happen, but it's the overall portfolio of outcomes that matters most. When you embrace a portfolio approach, you will be less inclined to dwell on individual losses, because you know that they are small parts of a much bigger picture.

BE ON THE LOOKOUT FOR REGRET MANIPULATION.

In chapter 8 we discussed how regret is sometimes used to manipulate decisions. If someone wants you to do A, but you are more inclined to do B, then his job is to make you believe that doing B instead of A will result in regret. It's accurate to describe this argument as a manipulation of "pre-regret," because you have not done anything yet to regret—the regret scenario is being painted for you as a worst possible outcome that you can avoid, if you choose, by making the "right" decision. At times, someone making you aware of impending regret is a good thing. If your friend is about to drive her car while drunk, you are doing her and others a great service by making her aware of what could happen and why the right decision is to give you the keys (and if she doesn't anyway, please take them). But when a salesperson at an electronics store is trying to sell you a product insurance plan by telling you how much you'll eventually regret not buying it—that's a manipulation.

REMEMBER HOW CHIMPS AND CHILDREN OVERCOME IMPULSIVITY.

Sometimes being impulsive is fine; other times it leads to trouble. Chimps and human kids, we saw in chapter 7, use a similar technique for overcoming impulsivity in the interest of getting a bigger reward later. By distracting themselves with toys, they are able to delay gratification longer and enjoy more candy than they would get if they took the first opportunity to grab it. Simple as that sounds, it's really a not-so-simple problem-solving strategy that human adults struggle with daily. But perhaps remembering that even a chimp can do it is inspiration enough to keep trying.

WORDS DIRECT PERCEPTION.

Here's an experiment to try out: Find a black marker and two paper bags; on one bag, write the word *Roses*. On the other, write the words *Chili Peppers*. Now put rose petals into each of the bags and close them up. Find a few people willing to lend their sniffing power to your cause and ask them to sniff each bag (making sure that they can read the labels but not see the contents of the bags). Then ask them to report on what they smell in each bag. That's a basic re-creation of a study that investigated whether the name of an object will affect what it smells like to the participant. In the study, most people mistook the rose petals in the second bag as chili peppers, even though the contents of the bags were identical.[9]

MODELING IS AS HUMAN AS BREATHING.

As discussed in chapter 14, each of us is a born imitator. Our brains are tuned to observe and re-create what we see in others, and this, like all habits of the happy brain, is both good and bad. On the upside, imitation plays a crucial role in learning. The downside is that imitation can spill over healthy boundaries and do us, and others, a world of harm. Two lessons come to mind. First, be careful about whose modeling is influencing you. Your brain has a difficult time distinguishing between good and bad lessons and will learn them with equal efficiency. Second, be careful about how you model for others. If you are a parent or guardian to young children, this lesson is especially important because kids' aptitude for imitation is exceptionally strong. Even when you think they aren't paying attention, you may be unknowingly modeling a behavior for them that you'll later regret.

LONELINESS AND CONFLICT ARE PARTNERS IN THE HUMAN BRAIN.

As we discussed in chapter 5, neuroscience research has found a convincing neural correlation between the experience of loneliness and an attraction to human conflict. Loneliness, by this definition, has nothing to do with how many people are physically nearby, and everything to do with *feeling* socially isolated. This research lends credence to the saying "Misery loves company," because people who feel socially isolated may be predisposed to seek out conflict. Everyone who has worked in an office with other people can relate to the finding.

ESCAPISM IS NOT MAGIC.

While it is true that certain forms of escapism have an inherently compulsive quality, it's also true that plenty of people play video games, play role-playing games, and participate in an endless list of other diversions without ever dancing too close to the compulsion pathway. Escapist diversions do not wield addictive magic over their patrons any more than mindless television shows make their viewers stupid. Caution is warranted because we know that some people are more prone to compulsive behavior than others, and when they bring those tendencies to an immersive role-playing game, for instance, we shouldn't be surprised when two weeks later they are spending ten hours a day playing it. The line between enjoyable diversion and compulsive behavior can be very thin and gray for some yet well-defined for others. We understand more now than we did even a decade ago about how the brain develops habits, compulsions, and addictions, but much is still unclear about these categories. Whether compulsive use of smartphones, for example, can rightly be called an "addiction" is still very much an open question, despite how loosely the term is used in much of the media. So as we move into an ever more saturated era of intense, immersive e-media, it's worth thinking about possible outcomes without veering into alarmism.

WORK IN LAYERS.

If you must attempt to multitask, try at least to strategically layer your work such that you are not pairing two resource-intensive things at the same time. For example, trying to reply to an email on your smartphone while driving is an exceedingly bad idea. Both tasks require too much mental energy, and it is impossible to evenly parse attention between them. If you are speaking to someone on the phone (while not driving), it may be possible to do something at the same time like scan a magazine article, because in most cases these tasks will not overpower each other. If you are attempting to truly concentrate on an article or book, then the situation changes and you are again imbalanced. Better yet, avoid multitasking altogether and instead try shift-tasking, which means that you work through one task and then shift attention to another, and then shift back after a while, or to something else altogether— like a train switching tracks. Arguably, this is still not an especially efficient way to complete projects, but the pace of our lives seldom allows for blocks of uninterrupted time to get things done. Shifting between competing priorities may be the best we can manage until a rare windfall of time becomes available.

Chapter 16

SHAKE YOUR MEANING MAKER

"Our obligation is to give meaning to life and in doing so to overcome the passive, indifferent life."
—ELIE WIESEL, *ESSAY ON INDIFFERENCE*

"You will never be happy if you continue to search for what happiness consists of. You will never live if you are looking for the meaning of life."
—ALBERT CAMUS, *YOUTHFUL WRITINGS*

I n closing, I have three thoughts I would like to leave you with, albeit delivered somewhat indirectly.

STOP HORSING AROUND

In a lesser-known section of Jonathan Swift's iconic tale *Gulliver's Travels*, Lemuel Gulliver encounters a race of beings resembling strikingly handsome horses. The Houyhnhnms (a name meaning "perfection of nature") are quintessentially rational, refined, and intelligent—and to Gulliver's beleaguered eyes, they are perfect. He views them in stark contrast to the Yahoos, a race of humanlike creatures ruled by the Houyhnhnms that are emotional, dirty, and stupid. Later, when Gulliver returns to live among humans, he can't help but see those around him as merely Yahoos with slightly higher social standards. He never recovers from the despair that this comparison brings, and he spends most of his time talking to the horses in his stable.

If perfection were attainable and a minority of humans finally achieved it, then, like Gulliver, most of us would compare ourselves to those models of perfection and despair that we fall short. Our deficiencies and flaws would become glaring symbols of our imperfection. Most of those around us (the other imperfects) would seem faulty as well, though with time the imperfect strata of society might coalesce. Eventually we would convince ourselves that the very fact that some humans have attained perfection means that many of us, given enough time and effort, can do so too, because the factors that differentiate perfection from imperfection would be identifiable and correctable. Soon people would write books identifying the problems that must be corrected for one to become perfect and offer systems for doing exactly that. If you follow the formula, then you will be able to leave the common strata of imperfects and join the ranks of the perfect minority.

I have a feeling that the sardonic Mr. Swift had something like this scenario in mind when he cast Gulliver into depression over humanity's flaws. We, of course, do not have any models of perfection to observe and compare ourselves to, but that doesn't stop us from comparing ourselves to an imagined ideal of perfection. An enormous portion of the media and entertainment industry is devoted to fostering this comparison and selling products to bridge the chasm between us imperfects and the flawless ideal. The number of systems for becoming the "perfect you" are legion. You only have to stroll through a bookstore, or go virtually anywhere online, to find more of them than you'll care to count. Unfortunately for us, our brains are susceptible to these messages.

That's frustrating—but because we are susceptible doesn't entail having to fall for it. Our brain is the most advanced imperfect wonder of nature on the planet, and perfection to any degree is not on the evolutionary docket. Better that we come to terms with our shortcomings, both hardwired and learned, and dance the awareness-action two-step to live more fulfilled lives.

MEANING, YOU ASK?

Typing the phrase "meaning of life" into a Google search bar turns up 71.3 million results (at least, that's the number as I write this; in 2011 that number was "only" 6.1 million). It's clearly a topic we think about quite a lot. If you scour through the list, most of the pages have something to say about "finding" the meaning of life. A great deal of these pages have a spiritual flavor, and nearly an equal amount offer guidelines and formulas for finding the "grail" of meaning. A smaller number, though still significant, are dark meanderings about the meaninglessness of existence.

I doubt there was ever a point in human history when meaning was not thought of as something to be found. We are the only existential animal—the only being on this planet with a mind able to look upon itself and ask, "Why?" If the answer to the question of meaning cannot be found within, we will search outside ourselves, and we have been doing just that for the larger part of our relatively short stay on Earth. The irony is that our brains evolved to make sense of our world—and they're rather good at meeting that challenge—but they routinely fail at making sense of *us*.

I would argue, alongside Wiesel and Camus, that the real problem is the question itself. Asking where meaning can be found is a diversion from the real challenge we humans face every day: to make meaning of our lives. This is a challenge only an existential animal can take on; it is our burden alone to answer questions about our world that go well beyond instinctual reaction and rudimentary learning. This is perhaps the greatest distinguishing feature of our minds—to make meaning of our experience and live out that meaning. Another way to state the last point is simply that this tremendous capacity for making meaning of our lives—unrivaled in the natural world—guides our behavior.

COGNITO FINITO

Wrestling with the stubborn tendencies of the happy brain is at times frustrating, exhausting, and even infuriating. We often find ourselves thinking and acting in ways that do not serve our best interests—though exactly what those interests are is rarely clear in the moment. We are subject to an array of seen and unseen influences, and in our more desperate moments it may seem as though our brains are conspiring with these influences against us. Living is, after all, messy business, and more often than not it is ambiguity rather than clarity filling our mind-space.

The final word, however, is still ours. We are the meaning makers—enabled by a brain more advanced than anything else on the planet, a brain that has brought us quite far, and will continue to push us forward. I hope this book has provided you with a few more clues and suggestions for understanding that incredible organ and for making meaning in your life.

SUGGESTED RESOURCES

In this section I offer a selection of sources that constitute an enormous amount of knowledge on all things mind and are well worth a knowledge-hungry reader's time to find and devour.

The Man Who Wasn't There: Investigations into the Strange New Science of the Self
    By Anil Ananthaswamy
    Dutton (2015)
    A truly elite science exploration, this book pushes further into the burgeoning understanding of how the brain creates the self, which in turn provides our linear self-narrative—the reason we think of ourselves as an "I." The self, by this definition, is a unifying leveler of our erratic neural natures—an adaptive way to bring order from chaos—and when this process goes off the rails, some odd and unnerving things can happen. The author uses this lens to examine Alzheimer's disease, schizophrenia, ecstatic epilepsy, depersonalization, Cotard's syndrome, and other conditions in which the self has become detached from its moorings.

Predictably Irrational: The Hidden Forces That Shape Our Decisions
    By Dan Ariely
    Harper Perennial (2010)
    Dan Ariely is a standard-bearer in behavioral science, and one of the most insightful thinkers writing on human irrationality today. His research has helped dispel the myth of "Homo economicus" and shone light on the unseen influences that affect every decision we make. If not

for Ariely's work, a book like the one you have in your hands would never have been written.

Beyond the Brain: How Body and Environment Shape Animal and Human Minds
By Louise Barrett
Princeton University Press (2011)
Louise Barrett asks us to put aside our "human-centered spectacles" and reconsider both human and animal cognition. Rather than focusing solely on the brain, she emphasizes that thinking and behavior are outcomes of an organism functioning as a whole. To make her argument, she delves deeply into cognitive science, comparative psychology, artificial intelligence, and a range of other topics.

Remembering: An Experimental and Social Study
By Sir Frederick Bartlett
Cambridge University Press (1932)
Bartlett was the first to assert that our memories are not recollections, but reconstructions. It would take many decades before his argument could be borne out by neuroscience research, but eventually his early work would prove well ahead of its time.

Can't Just Stop: An Investigation of Compulsions
By Sharon Begley
Simon and Schuster (2017)
Begley's book attempts to understand the distinction between habits, compulsions, and addictions. While there's now a wealth of science addressing the question, none of it is entirely clear-cut, despite the conclusions we see batted around popular media. This book tends to the murky areas by trying to make sense of what is known, while not forgetting that much is still unknown. It's a solid entry in an ongoing discussion.

The Meme Machine
By Susan Blackmore
Oxford University Press (2001)
Susan Blackmore is best known for endorsing meme theory—that ideas are replicated from mind to mind in a way similar to gene replication at the biological level. Her opus, and still one of the most highly regarded books in the field, is *The Meme Machine*, in which she lays out the pieces of meme theory in an accessible and relevant way. For anyone intrigued by why certain ideas inexplicably pop up across the globe with no apparent connection, this is a worthwhile read that will leave you with much to think about and explore.

Against Empathy: The Case for Rational Compassion
By Paul Bloom
Ecco (2016)
Is empathy the same as compassion? Paul Bloom says it isn't, and he's marshaled convincing reasons for you to believe him. You have to admire a book that takes a tough stand against an entrenched position, knowing full well the blowback it's going to receive. Bloom's argument that our hyper-emphasis on empathy is actually hurting us is controversial, no doubt, and you might bristle hearing it. But Bloom is such a persuasive apologist for his position—a position supported by more research than most realize—it's worth reading to at least expand your perspective, and quite possibly come away thinking differently about a topic that's a giant magnet for assumptions.

Play Anything: The Pleasure of Limits, the Uses of Boredom, and the Secret of Games
By Ian Bogost
Basic Books (2016)
This book falls in the "Tackle Life's Challenges Like a Game" category, a thesis that's gaining momentum, but this volume goes deeper than

most via an enlightening discussion of the role of limits in both games and life. The author strikes me as equal parts philosopher and savant game enthusiast—a systems thinker with a penchant for high-score formulas—and I'm glad he wrote *Play Anything*, because it's causing me to look at problems in a different way.

Sway: The Irresistible Pull of Irrational Behavior
   By Ori Brafman and Rom Brafman
   Crown Business (2008)
   Ori and Rom Brafman are part of a cadre of excellent writers taking on the human rationality shibboleth. Along with Dan Ariely, they are in part responsible for a new view of why we humans think as we think and act as we act—and the explanation is hardly rational.

Suspicious Minds: Why We Believe Conspiracy Theories
   By Rob Brotherton
   Bloomsbury Sigma (2015)
   Brotherton provides historical and cultural context for understanding the draw of conspiracy thinking, and also delivers a solidly research-based explanation. Conspiracy thinking isn't exclusively about cognitive biases like confirmation bias, proportionality bias, and agency misattribution (although they play a significant role), it's really the predictable outcome of brains fueled by that immensely powerful tonic called belief. Since that's how all of our brains work, conspiracy thinking is best understood as inherent in our natures, though for a variety of reasons it takes fuller bloom in some more than in others.

On Being Certain: Believing You Are Right Even When You're Not
   By Robert Burton, MD
   St. Martin's (2008)
   Robert Burton effectively argues that many of our most stubborn truth stances are not really about *being* right but about *feeling* right. He

posits that neural connections between a thought and the sensation of being correct strengthen over time because the brain experiences the sensation as a reward. The longer the reward is received, the greater the reinforcement of the connection. Burton's book explains this and related issues cogently and conversationally, without glossing over evidence-based points that are often left out of less substantial books.

Loneliness: Human Nature and the Need for Social Connection
    By John T. Cacioppo and William Patrick
    W. W. Norton (2008)
    In this quintessential book on human loneliness, Cacioppo and Patrick tell us that being lonely has little to do with how many people are physically around us, and much more to do with falling short of getting what we need from our relationships. Their message is particularly relevant in the age of social networking, when someone can have a thousand "friends" online and still feel alone in the world.

Multiplicity: The New Science of Personality, Identity, and the Self
    By Rita Carter
    Little, Brown (2008)
    Why does each of us think of ourselves as an individual "I" when in truth we are several different selves during the course of any given day? Rita Carter tackles that question in what I consider to be the benchmark book for multiple-selves theory. Carter, a master explainer of complex topics, convincingly argues that what we consider to be the self is not merely one thing, though we operate under an illusion that it is.

The Invisible Gorilla: And Other Ways Our Intuitions Deceive Us
    By Christopher Chabris and Daniel Simons
    Crown (2010)
    Chabris and Simons are the creators of the famous "invisible gorilla" experiment, video versions of which can be found all over the Internet.

Their original experiment is, in my opinion, still the most convincing illustration of how we can fail to "see" the obvious while thinking our perception is seamless. In this book, the researchers elaborate on the gorilla experiment, along with many others, and explain why we are prone to missing what's right before our eyes, and why trusting intuition can be risky business.

Harnessed: How Language and Music Mimicked Nature and Transformed Ape to Man
   By Mark Changizi
   BenBella Books (2011)
   Mark Changizi is a neuroscientist more interested in "Why" questions than in "How" questions. In his previous book, *The Vision Revolution*, he offered a new way of thinking about why human perception evolved as it did, and in this book he takes on the even larger question of what factors contributed to the massive evolutionary step that resulted in modern humans. Changizi is always thorough; and his insights are always provocative.

Algorithms to Live By: The Computer Science of Human Decisions
   By Brian Christian and Tom Griffiths
   Henry Holt (2016)
   Don't bother organizing your emails or your office desk. You can also stop searching for a time management system or an ideal scheduling plan. And when you're looking for somewhere to eat, don't spend too much time considering new places versus your old standbys. The reasons why (and reasons for many more suggestions) are best revealed by understanding the algorithms that exert influence on our daily lives—so argue the authors of this surprisingly useful book that travels from computer science to human decision making with more fluidity than you might guess.

Influence: The Psychology of Persuasion
By Robert Cialdini, PhD
Harper Paperbacks (2006)
Considered by many to be the defining book on the power of influence, *Influence* will change the way you think about negotiations, sales, and every other scenario in which persuasion is the overriding dynamic. Fifty years from now, Cialdini's book will continue to be a milestone work in the field.

Self Comes to Mind: Constructing the Conscious Brain
By Antonio Damasio
Pantheon (2010)
Many have tried to explain consciousness, but it remains the most vexing topic in cognitive science (and possibly all of science). Antonio Damasio, one of the world's leading neuroscientists, adds his voice and estimable contribution to the discussion. An important read for anyone interested in following the latest evidence-based thinking on consciousness.

Freedom Evolves
By Daniel C. Dennett
Viking Adult (2003)
Philosopher Daniel C. Dennett combines evolutionary biology, cognitive science, economics, and several other fields to make a strong argument that ethics evolve. As always, he manages to say something new and challenging about an argument many others have attempted to cover, but none so skillfully as Dennett.

The Brain That Changes Itself: Stories of Personal Triumph from the Frontiers of Brain Science
By Norman Doidge, MD
Penguin (2007)
Norman Doidge wrote one of the first popular overviews of brain-plasticity research, and it still remains among the best on the shelf

addressing this exciting topic. The brilliance of Doidge's work is that he makes extremely complicated neuroscience concepts relevant in ways you wouldn't expect. I continue to go back to Doidge for elaboration on brain plasticity, even as a flood of neuroscience books are published each year.

Wrong: Why Experts Keep Failing Us and How to Know When Not to Trust Them
By David H. Freedman
Little, Brown (2010)
David H. Freedman takes on the mythos of expertise in this tightly argued exposition. In his view, we are over-reliant on experts and reticent to challenge them because of their positions—while as a society we suffer from their often-poor conclusions. His argument is refreshing and provides needed balance in the debate over the value of expertise.

The Checklist Manifesto: How to Get Things Right
By Atul Gawande
Picador (2009)
Atul Gawande may prove to be one of the most influential thinkers in recent memory, though the subject he writes about is deceptively simple: checklists. But when you read just how important this simple tool has been to the healthcare industry—and how many lives have been saved by its use—you'll probably agree that Gawande's voice is one worthy of attention.

Human
By Michael S. Gazzaniga
HarperCollins (2008)
Gazzaniga is among a handful of scientists responsible for advancing the new wave of cognitive-science research and, in the process, changed how we think about our minds. In *Human*, he takes a comprehensive approach to what being human is all about. A splendid contribution from a pioneer in the field.

Stumbling on Happiness
By Daniel Gilbert
Vintage Books (2007)
I personally credit Daniel Gilbert for being among the most proficient at making difficult topics born from complicated research accessible to a broad audience—and doing so without compromising intellectual integrity. *Stumbling on Happiness* has deservedly risen to the upper echelon of popular psychology books and more than merits being read. Gilbert's premise is encapsulated in the question: Do you really know what makes you happy? If you think you do, all the more reason to read this book.

Bad Science: Quacks, Hacks, and Big Pharma Flacks
By Ben Goldacre
Faber and Faber (2008)
Ben Goldacre is a thorn in the side of pseudoscience, and I applaud him for it. In *Bad Science*, he methodically deconstructs the flawed arguments of quacks and cajolers, and in so doing provides his readers with much-needed ammunition against those who would deceive us in the name of "science."

The Happiness Hypothesis: Finding Modern Truth in Ancient Wisdom
By Jonathan Haidt
Basic Books (2006)
Jonathan Haidt works at the leading edge of research on morality and the emotional underpinnings of belief. *The Happiness Hypothesis* brings together findings from the research literature with ten influential ideas drawn from centuries of learned wisdom. I rank it among the most insightful works of psychological synthesis written to date.

Why We Make Mistakes
By Joseph T. Hallinan
Broadway Books (2009)

Pulitzer Prize–winning journalist Joseph T. Hallinan is among the best writers at making the complex understandable, and in this conversational work he does exactly that. If you have ever wondered why you fail to see the beer you're looking for in the refrigerator, though it's literally right in front of you, you'll enjoy this witty and enlightening book.

Our Kind: Who We Are, Where We Came from, Where We Are Going
 By Marvin Harris
 Harper Perennial (1998)
 Marvin Harris was an anthropologist whose writings influenced a generation of up-and-coming social-science thinkers, and his many books continue to influence readers today. In *Our Kind*, readers receive the best of Harris across a gamut of topics, in essays ranging from "The Need to Be Loved" to "Why We Became Religious." To say it is anything less than an intellectual feast would be an understatement.

The Moral Landscape
 By Sam Harris
 Free Press (2010)
 Sam Harris is a writer at the center of controversy, but what you may not know is that he is also a neuroscientist who studies the neural infrastructure of belief. *The Moral Landscape* is his attempt to show that it is possible to live a moral life without the influence of religion. I rate it as an important book, and likely a pathbreaking work that others will follow.

On Second Thought: Outsmarting Your Mind's Hard-Wired Habits
 By Wray Herbert
 Crown Publishing Group (2010)
 Wray Herbert was my model for how to write about psychology topics when I first started as a science writer. His clear, jargon-free prose and his attention to the finer details of research makes him a must-read for anyone interested in the topics I've discussed in this book. *On*

Second Thought is a masterful exposition on the influence of heuristics on thought. I give it my highest recommendation.

SuperSense: Why We Believe the Unbelievable
By Bruce Hood
HarperOne (2009)
Cognitive psychologist Bruce Hood argues that we are born with a "supersense"—a predisposition to believe in supernatural explanations when other explanations don't satisfy our brain's need for comprehensible meaning. Hood's points are well researched and his style lends well to being understood by readers without a background in cognitive science.

The Scientific American Brave New Brain
By Judith Horstman
Jossey-Bass (2010)
Bringing together recent advances in cognitive science in a digestible format, Judith Horstman's book is essential reading for those wanting to know more about the latest discoveries in this exciting field. The book not only breaks down recent findings to inform the reader about what we currently know, but also projects into the future to give a sense of where cognitive science may be going next.

The Owner's Manual for the Brain: Everyday Applications from Mind-Brain Research
By Pierce J. Howard, PhD
Bard (2006)
When it comes to exhaustive overviews of cognitive science, few compare to Pierce J. Howard's monumental tome. If a reader is searching for a place to begin reading about the brain, this is an excellent choice. Howard's writing is accessible to a general audience and grounded in evidence-based propositions about how the brain works.

The Art of Choosing
   By Sheena Iyengar
   Hachette Book Group (2010)
   Sheena Iyengar is a business professor who has conducted extensive research on the factors influencing decisions. In *The Art of Choosing* she asks: How much control do we really have over the choices we make? By weaving together a variety of relevant examples with credible research findings, she leads the reader closer to an answer while opening new questions along the way.

Meaning of Truth
   By William James
   Public Domain Books (2004)
   A classic work by the godfather of modern psychology. Along with all of his works, this one deserves an esteemed place on your bookshelf.

The Rough Guide to Psychology
   By Christian Jarrett
   Rough Guides (2011)
   Christian Jarrett is the writer behind the highly regarded *British Psychological Society Research Blog*, where he makes the latest psychological research accessible to a large audience. This book crystallizes the best of his work and advances the goal of opening the world of psychology to readers who may have otherwise never found it.

Born to Be Good: The Science of a Meaningful Life
   By Dacher Keltner
   W. W. Norton (2009)
   Keltner's book is among the most credible evidence-based expositions on positive psychology written to date. Rather than "survival of the fittest," Keltner argues that we are evolutionarily wired for kindness, altruism, and positive emotional response. Drawing on sources from

Darwin to modern-day neuroscience, Keltner makes a compelling case that the key to our success as a species resides less in our survival instincts and more in our capacity for connection with other human beings.

The Confidence Game: Why We Fall For It . . . Every Time
By Maria Konnikova
Viking (2016)
Why are we so easily and repeatedly fooled even when we've fallen for the same con before? If you want to know, and really want to understand the backstory of your gullibility, this is the book to read. It's an exploration of how cons work and why we fall for them—a deconstruction of psychological dynamics, tactics, ploys, and susceptibilities illuminated by an excellent communicator. Whether it's a parking lot grifter, a pyramid-scheme huckster, or a political manipulator, this book explains how and why their deception so often works. Read it and maybe you'll be better prepared the next time someone starts a sentence with, "Imagine this . . ." or ends one with ". . . believe me."

Why Everyone (Else) Is a Hypocrite: Evolution and the Modular Mind
By Robert Kurzban
Princeton University Press (2011)
Robert Kurzban is an evolutionary psychologist and an engaging writer. In this book, he throws light on the experience of contradiction in the human mind and effectively argues that all of us contradict ourselves, and that doing so is fundamental to how our brains work.

Anxious: Using the Brain to Understand and Treat Fear and Anxiety
By Joseph LeDoux
Penguin (2015)
Quite possibly the most trenchant exploration of the causes of anxiety ever published as a nonacademic book, Le Doux's work is a treatise on a condition that affects at least 20 percent of the population. Le Doux's lab

at New York University is a hub of anxiety and fear-disorder research, and he approaches the topic as a veteran researcher. He argues that anxiety disorders are maladaptive responses closely tied to the brain's "defensive circuitry" and how the brain processes uncertainty. He eventually brings the discussion around to cutting-edge approaches that are showing progress in treating this notoriously hard-to-treat condition.

The Biology of Desire: Why Addiction Is Not a Disease
    By Marc Lewis, PhD
    Public Affairs (2015)
    Marc Lewis is a neuroscientist and recovering addict, and his voice in addiction studies is distinct. In this book he takes a somewhat-controversial position, arguing that addiction isn't a disease but "an unfortunate outcome of a normal neural mechanism that evolved because it was useful." Addiction is, in Lewis' view, a "goal-directed activity" fueled by desire, and our brain circuitry facilitates the pursuit of goals with powerful neural chemicals like dopamine. Unfortunately, the same process that allows us to pursue beneficial goals also enables addiction, to chemicals like drugs and alcohol and behaviors like gambling and sex.

Cure: A Journey into the Science of Mind Over Body
    By Jo Marchant
    Crown (2016)
    We're in a period of intense interest in the ways the mind influences the body. The placebo effect, for example, publicly surfaced seventy or so years ago, but the science behind why it works is only now becoming clearer. That's exciting, but the downside of this elevated interest is that frauds and over-staters have grabbed onto the mind–body connection and flooded the public with claims ranging from silly to irresponsible (i.e., positive thinking can cure cancer). This is an important book that cuts through the nonsense and clarifies the science.

Kluge: The Haphazard Construction of the Human Mind
  By Gary Marcus
  Houghton Mifflin Company (2008)
  In this short but well-written book, professor of psychology Gary Marcus explains that the human brain is not a well-oiled machine but a "kluge," an engineering term for something put together haphazardly that somehow still works. For any reader looking for a solid overview of cognitive bias in a quick-to-read package, this book is a great choice.

Your Brain Is (Almost) Perfect: How We Make Decisions
  By Read Montague
  Plume (2007)
  Read Montague was one of the first authors to write about the new advances in cognitive science that illuminate what is going on in our brains when we make a decision. His book is still a mainstay for those interested in the science of decision making.

Personality: What Makes You the Way You Are
  By Daniel Nettle
  Oxford University Press (2007)
  Quite possibly the best short book on human personality I've read, Nettle's take on the topic is succinct and well argued. If you are interested to know more about why you are the way you are, this is an excellent, fast read to start with.

Mindware: Tools for Smart Thinking
  By Richard E. Nisbett
  Farrar, Straus and Giroux (2015)
  Nisbett's book is a toolbox for better thinking. He uses an interdisciplinary approach that borrows from different fields of science to support his premise that scientific and philosophical reasoning can be taught. That may sound like a lofty goal, but the material is relatable and accessible. It's a timely read designed to enable slicing through the detritus of information overload.

Do Gentlemen Really Prefer Blondes? Bodies, Behavior, and Brains: The Science behind Sex, Love, and Attraction
   By Jena Pincott
   Delacorte (2008)
   Jena Pincott is a science writer with a superb skill for communicating with broad audiences. In *Do Gentlemen Really Prefer Blondes?* she presents an array of topics that many of us have thought about but few of us talk about. Her approach is both funny and trenchant, and you will come away from her book feeling like you know 100 percent more about sex, love, and attraction than you did before.

Drive: The Surprising Truth about What Motivates Us
   By Daniel H. Pink
   Riverhead Books (2009)
   Run-of-the-mill discussions of motivation can't compare with Pink's novel approach to the subject. Author of the bestselling *A Whole New Mind*, Pink is a master at introducing readers to a different perspective, and his ideas are well worth considering.

Brains: How They Seem to Work
   By Dale Purves
   FT Press (2010)
   I especially like the title to Dale Purves's book, which rather straightforwardly implies that we don't know for certain how our brains work. This is no doubt true, but Purves manages to make several worthwhile contributions within the parameters of "seem" that merit reading.

Wellbeing: The Five Essential Elements
   By Tom Rath and Jim Harter
   Gallup (2010)
   The Gallup organization has spearheaded quite a bit of research on strengths, talents, motivation, and personality. I've enjoyed several books

from Gallup, and *Wellbeing* certainly deserves to be on your reading list if you want to know what survey research has to say about what makes us feel fulfilled. Unlike many other books on similar topics, this one is grounded in well-documented findings.

The Conquest of Happiness
By Bertrand Russell
Liveright (1971)
Bertrand Russell was, and continues to be, one of the most influential Western philosophers of the past century. In this book, he takes a break from his work as a logician and turns his powerful perception on a topic that affects us all. The result is, in my opinion, one of the most compelling books ever written on happiness.

Why Zebras Don't Get Ulcers
By Robert M. Sapolsky
Holt Paperbacks (2010)
Robert M. Sapolsky is one of the world's leading experts on how stress affects organisms. More important, he is one of the most engaging writers and speakers on topics touching upon the human mind. This book is an engaging read that will leave you feeling differently about several assumptions we usually take for granted.

Behave: The Biology of Humans at Our Best and Our Worst
By Robert M. Sapolsky
Penguin (2017)
It's difficult to imagine a more comprehensive book taking on the complex "whys" of human thought and behavior than this book, and it's equally hard to imagine a better communicator of the content than Robert Sapolsky. From the brain's reward system to the neurochemical underpinnings of morality, this masterwork covers more ground, with more grounded detail, in over seven hundred pages, than any other single book I can recommend.

SUGGESTED RESOURCES

The Paradox of Choice: Why More Is Less
  By Barry Schwartz
  Harper Perennial (2004)
  Every so often a book is written that nails its topic; Schwartz's book is definitely one of them. When you start reading, you won't want to stop until you've digested it all. Unlike an academic approach to the topic of choice, Schwartz brings it down to a practical level to show how endless options are not making us more satisfied but actually do much the opposite.

Science Friction: Where the Known Meets the Unknown
  By Michael Shermer
  Owl Books (2005)
  Founder of *Skeptic* magazine Michael Shermer is expert at debunking pseudoscience claims and showing what's really going on behind the curtain. This book is an entertaining uncloaking of deceptions like "cold reading," a skill Shermer learned and successfully used himself to demonstrate that it is, indeed, a sham.

iBrain: Surviving the Technological Alteration of the Modern Mind
  By Gary Small, MD
  Collins Living (2008)
  iBrain is a foundational book for understanding how the Internet and other new media technologies are affecting our brains. Neither alarmist nor indulgent, Small's argument is well grounded and pragmatic. I recommend it for anyone interested in knowing more about how our minds interact with the new media world we are only just starting to live in.

The Empathy Gap: Building Bridges to the Good Life and the Good Society
  By J. D. Trout
  Viking (2009)
  J. D. Trout brings together lines of thought from philosophy and psy-

chology to address what he calls the *empathy gap* and its consequences. Trout brings to bear an impressive number of sources, from antiquity through modernity, to make the argument that only empathy can produce the lasting change our society needs to thrive, and we have to be willing to vault the empathy gap if we want a "good life and good society."

Special Section 2

OF TECHNOLOGY AND REWARDS

For whatever else we humans are, we're unquestionably lovers of our technologies—the continuously changing creations of our predicting and pattern-detecting brains. Since the first edition of this book (in 2011), smartphones—as a prime example of a technology that captivates us—have become so integrated into our lives that it's getting harder to imagine a time without them. Why we're so enmeshed with our smartphones is a story long in the making, and it says a lot about our brains' patterns of reward seeking. Smartphones are the most accessible reward-delivery devices to date. They give our brains exactly what "makes them happy." Other technologies are developing that could push reward delivery even further by making accessibility nearly automatic.

Our relationship with these technologies is well worth examining. But first, we need to have a broader discussion about how our brains respond to rewards, and that first requires that we talk a little about learning.

DANGEROUS LEARNING (COURTESY OF THE DOUBLE AGENT IN YOUR HEAD)

One of the hardest things for us humans to realize is that how our brains are natively wired to learn—as amazing a capability as this is—also predisposes us to developing problematic behaviors. The ironic twist is that the same neural apparatus that sets the stage for these problems is what allows us to achieve, succeed, and grow beyond our imaginations. In other words, the brain that makes us is the same brain that can undo us, like a double agent as committed to one allegiance as the other, always ready to switch sides depending on the circumstances.

The main driver behind all of this is our remarkable and unparalleled capacity to learn. We come into this world like sponges soaking up the details of our new realities, and with time that infantile spongy genius morphs into a more refined but no less incredible ability that we carry with us into adulthood and for the rest of our lives. We are learning all of the time, and that learning moves along patterns that influence who we are.

To place this in a one-sentence package: *Learned patterns of thought yield behavior, for both the good and the bad.*

The gambler learns a pattern of thinking that yields a particular behavior. That thinking pattern is fueled by engaging with the behavior it yields.

The health-minded gym enthusiast also learns a pattern of thinking that yields a behavior, and that thinking pattern is also fueled by engagement with the behavior it yields.

To the brain, these thinking patterns are different in content, though not in structure. The same underlying neural structure supports the development and reinforcement of different behaviors—one that may lead to a potentially self-destructive lifestyle, another which may lead to improved physical health. The "fuel" is the same, the neural structures are the same—it's the behavior and its consequences that are different.

And so it goes for us and our double-agent brains. With each learned pattern of thinking, we experience an internal, fueled momentum along a neural structure that will result in a behavioral outcome. Sometime these behaviors barely matter, other times they'll matter more than anything else.

The problem is that a slew of thinking patterns and their resulting behaviors are not in our best interest, or at least may move in a direction that will ultimately prove wrongheaded or dangerous. What at first seems innocuous can eventually undermine our health and sanity. But because our brains suffer a sort of directional blindness when it comes to choosing thinking patterns, we are constantly faced with the paradox embodied in the title of this book. And the challenge is always about how to distinguish between negative and positive patterns, even when we're unable to predict where the patterns will lead us.

This story is all about learning, and learning is about the pursuit of "rewards."

REVISITING THE REWARD CENTER

In chapter 5, "Immersion and the Great Escape," I gave a more detailed description of the brain's reward center and the effects of various types of rewards "imprinting" on its neural circuitry (it's worth scanning through it again if you want a little more context for this section). Here's a snippet reminder from that chapter:

> This reward center (called the *mesolimbic reward center*), while indispensable to us, is not unlike an unprotected power grid in that it can be hijacked and tapped into from external forces. These forces make use of the same reward circuitry that benefits us in so many ways, and this circuitry (called *incentive salience circuitry* or ISC) adaptively responds just as it does to accommodate beneficial rewards. The problem is that the new rewards imprinting the ISC are generally not beneficial. But our brains suffer a sort of reward-distinction blindness, and new imprints are integrated into the grid.

Learning is about the pursuit of rewards, and it's a tremendously powerful capability, because our brains are structured to adaptively respond to new inputs on our reward circuitry. Whether we are learning what will ultimately be a beneficial or destructive behavior, our brains take it on with the full force of a system structured to adapt and learn better than anything else on the planet.

It doesn't take much detective work to see how these dynamics play out with our social media technologies, and the ultimate reward-delivery system: smartphones. We are continuously learning new reward pursuits via these technologies, and our brains are tirelessly integrating reward imprints that in turn fuel our engagement with the technologies. This cycle keeps us responding—keeps us checking, clicking, opening, replying, and on and on.

We always have to remember: *our brains are structured to respond.*

TRIGGERS, OFFERINGS, AND ACCESS POINTS

Let's break all of this down a bit more.

How technology interacts with the reward center is largely a story of *triggers, offerings, and access points.* The triggers are catalysts for reward pursuit. Access points are the entryways to engage the pursuit. Offerings are what we're pursuing. To our brains, triggers, offerings, and access points are essentially all one flavor—the topic is secondary. What matters is the pattern that's engaged by the reward center once a reward pursuit has been triggered, and the level of intensity of the pursuit.

For example, for a gambler, the trigger could be the thought of getting in a game (maybe from someone bringing it up in conversation, watching a friend play, seeing something online, or just thinking about the feeling of playing). The access point is the door to engaging the pursuit, which might be a gambling website or app, booking a trip to Vegas, or finding a basement game somewhere. The offering is whatever the game is, whatever form the gambling will take. Again, the specifics are secondary. What matters is that in our immersive digital world, both triggers and access points come at us from all directions, all of the time. If you have a compulsive gambling problem, and the only access points available are organizing expensive trips or finding covert games in town—or other logistically intensive efforts—you have a better shot of controlling the behavior, as there are more barriers to entry. But in the digital world, there are so many ways to be triggered and so many easy access points, you can hardly avoid them. They're available all of the time, and your brain is structured to respond to them.

Again, that's a critical part of this discussion that all of us need to get very, very clear about: *our brains are structured to respond.* So those designing and marketing all sorts of offerings focus on the ways the brain responds and imbue their creations with ever sharper, more focused attributes that our brains crave. It's simply a matter of selling to the need. If you're selling groceries, the best time to get someone in your store is after work, before dinner because they're tired and starving; hunger drives impulses to put more stuff in their cart, while their self-control muscles are too depleted to head off the impulse.

Likewise, if you're selling compulsive offerings of any sort, the best time to engage the reward system is when compulsions are driving a need to find triggers to kick off reward pursuits to capture offerings. If you can design an app that provides triggers and easy access points leading to compulsively engaging offerings, you'll pull brains in—lots of them. The profit motive to do so is ever present, and it fuels the design and marketing of these offerings. Again, we must never forget: *our brains are structured to respond.*

And to really rev the reward engine, it's essential to trigger the brain's inbuilt penchant for stability with doses of exactly the opposite, discussed next.

THE UNRELENTING POWER OF MAYBE

As we've seen throughout this book, the human brain is inclined toward certainty, stability, and consistency. Unpredictability and uncertainty pose challenges we're not altogether proficient at handling. One reason for this is the neurochemical cascade produced when those challenges take center stage, particularly when they're associated with a reward, whether tangible or virtual.

What we know is that when uncertainty and unpredictability are introduced during the pursuit of rewards, the human brain explodes with the neurotransmitter dopamine—the chemical fuel of our brain's reward system. That is what Robert Sapolsky, professor of biology and neurology at Stanford University, refers to as the "addictive power of MAYBE."[1] The same holds true for other species—the exact same neurochemistry is at work. And it throws us humans into a habitual maelstrom.

Studies with monkeys show how this works. In one especially well-crafted experiment, monkeys were trained to press a lever ten times to get a tasty reward. But first, a light went on to signal the session was starting. When the signal light went on, the monkey did the work—pressing the lever—and then there was a delay before he received a reward (raisins).

Before the study, researchers believed dopamine levels would increase when the reward was received. That proved to be incorrect. What actually happened is that dopamine levels increased when the *signal* indicating that the session was about to begin went on.[2]

"Dopamine isn't about pleasure, it's about the anticipation of pleasure," commented Dr. Sapolsky while elaborating on the study.[3]

When the researchers changed the study such that the monkey received a reward only 50 percent of the time, dopamine levels skyrocketed even more. Introducing uncertainty of reward causes dopamine to pour out like a waterfall over the brain's reward circuitry. That's the addictive power of MAYBE.

To put a finer point on this: we can quantify the power of anticipation of reward to a very close degree, and *that means it can be manipulated*. And it is manipulated, all of the time. Our brains respond to this manipulation because the human brain—like that of other species—is structured to explode with desire when anticipation of reward becomes variable. Whether that variable is up or down—greater or lesser chances of getting whatever it is we're pursuing—our brains are structured to respond, and that's exactly what they do.

It's often said that an addict's true high doesn't start when a drug is ingested, it actually starts much earlier, from beginning the process of craving the drug, then finding and buying it, and then finally using it to get the chemical reward. The high doesn't start when the chemical hits the bloodstream, it starts when the process begins.

The same is true for all of us, for all manner of rewards, chemical or behavioral. Just as the monkey's dopamine explosion started when the light went on to signal that the session was about to begin, ours is triggered when the "signal" for whatever reward pursuit we're involved with begins. This is happening to varying degrees in the digital universe all of the time. From the spikes felt while getting intermittent signals from our smartphones, to the zing of playing well-designed video games, to the cascade experienced when gambling or looking for porn, to name just a few examples.

Consider this dynamic particularly when it comes to virtual communication. By definition, there's more "maybe" baked into our virtual existences. If you are communicating with someone online, you can easily be pulled into a vortex of maybes about what the other person is really thinking or doing. The questions also apply to in-person situations, of course, but the virtual realm adds a dimension of uncertainty that makes speculation of the other's intentions, thoughts, and actions compulsively compelling. If you're not especially mindful (and let's be honest, most of us are not), it doesn't take much to give yourself over to endless conjecture—a compulsion to try to predict what's happening at the other end of the communication. And since in most cases those questions can't be answered with anything close to concrete satisfaction, the power of maybe has free rein to manipulate our maybe-prone brains into compulsively hanging on every word of every text and email, every photo and ad hoc note.

Again, we must remember that our brains are structured exactly that way, and virtual interaction wasn't something natural evolution prepared us to manage. We are working with a handicap in the technocultures our brains have created.

The brain wants stability, certainty, and predictability to stay level. When uncertainty is introduced, dopamine saturates the reward center, and the more that happens, the less level is our reality. If it happens chronically, such that a pattern of dopamine skyrocketing occurs for particular reward pursuits (whether behaviors or other), we move into the realms of compulsion and addiction.

MORE DYNAMICS OF TECHNOLOGY AND REWARDS

This brings us back to the question of technology, and specifically the reward-delivery systems that have in short order become indispensable to us, with smartphones high on the list. We've talked about the role of learning, the reward center, and variability. A few other topics are also central to this discussion.

Anticipation. As discussed earlier, the reward system is tied to anticipation of rewards, which makes the period between thinking of a reward (the "signal") and actually receiving it very critical. The ongoing series of intermittent, variable rewards that smartphones serve up are tantalizing to the brain because they so effectively leverage this dynamic. We're forever anticipating the next little reward, and then the next and the next. This works right along with the "power of maybe" because anticipation has uncertainty baked into it. We're never quite sure what's coming next, or if it's coming.

What really charges this up is the meta-anticipation of hidden gems of rewards arriving in the mix. For every, say, fifty or so texts, posts, pictures, or whatever, just a few are going to really deliver the goods we want. We are neurochemically trained to drill down for these rewards, through however much mediocrity and nonsense is in the way.

To elicit more rewards, we contribute to the reward stream by making our own posts and waiting for a response. Every time we post something, we're starting an anticipation stopwatch that works on the same principle. The jolt is even more profound because it's coming from a reward process we initiated.

Anxiety. There's a debate about what model best describes why we get hooked on technologies. Some think the addiction model fits best, but the problem with that argument is that addictions operate with a distinct pleasure principle: someone starts using a drug to get the euphoria at the other end, whether for pain relief or otherwise. Over time the drug can't deliver the same way, so more and more is taken to get the effect. It's the same for behavioral addictions like gambling—the gambler wants the high that comes from taking the risk, but the risk has to get bigger over time to deliver the same feeling.

What may better fit compulsive technology use is the anxiety model. It's not purely a desire for pleasure but also a response to heightened stress and anxiety that keeps us clicking. We're not talking about overt, panic-level anxiety; anxiety is insidious precisely because it's usually not overt, but more like a charged hum in the background that seeps into our thoughts and actions.

As author Sharon Begley notes in her book on compulsive behavior, *Can't Just Stop*, "By making us feel we are always connected to the world, [smartphones] alleviate the anxiety that otherwise floods into us from feeling alone and untethered."[4] Research on the topic backs up that argument, showing that compulsive smartphone use is tied closely to anxiety, even more so for those with anxiety disorders, which happens to be an enormous and growing segment of the population. The same conclusions have been reached by research on problematic Internet use in general: anxiety is the strongest precursor.

One component of that anxiety is that we don't like being alone with our thoughts. Begley discusses a study showing how for many people receiving a mild electric shock is preferable to doing nothing in a room for fifteen minutes.[5] That makes sense when you consider how our smartphones now fill all of the empty time slots in our days, times we'd otherwise be alone to think. The technology fills the empty space available and then creeps into spaces already filled with things like conversations with people in front of us.

FOMO. Compounding that anxiety, and fueling it every step of the way, is the constant fear that we'll miss the important rewards. While the term FOMO (fear of missing out) may have started as Urban Dictionary slang, it's an on-the-mark description of a key brain dynamic. Some psychologists have defined it as "a desire to stay continually connected with what others are doing"—because it's exactly "what others are doing," whoever they may be, that drives the reward machine.[6] If we're not connected, the rewards pass us by and we feel a sense of loss. And we also miss the opportunity to contribute to the reward stream and anticipate the rewards we'll receive in response.

Withdrawal. When you combine that sense of loss with anxiety, you have the sensation that comes from feeling disconnected, and it's a form of withdrawal. Here again we have to be careful with definitions, because withdrawal is mostly associated with the addiction model. In the anxiety model, we're struck by more and stronger anxiety from feeling disconnected, via the digital conduit that keeps us tethered to the world. It's not the same as addiction withdrawal (especially in chemical addictions), but there are similar elements. When the brain has engaged and established a habit that alleviates anxiety, and that habit is disrupted, then the predictable result is even more anxiety. Does anyone want to feel that? No, so the urge to keep our smartphone nearby at all times is a hedge against feeling anything like it.

SO WHERE DOES THIS LEAVE US?

The relationship between technology and the brain is a complex story, of course, as all brain stories are, but the dynamics we just walked through constitute much of the drama. Knowing all of this, you can see why those with a vested interest in keeping us hooked pay close attention to how these dynamics can be manipulated.

We have to remember that our brains are structured to respond. We're carrying around elegantly well-designed reward-delivery systems with us almost all of the time. Access points are always available, as are the multitude of offerings for anything we can imagine. Our challenge is how to manage our response, with brains structured to respond.

I think the bridge to optimism in this case is built with knowledge combined with techniques for managing our attention. Once you know what's going on in your brain when faced with compulsively compelling offerings, you have a starting point for action, and that action is enabled by finding reliable techniques for managing your attention. Because attention, after all, is where this begins—you can't be captivated by something unless it triggers your attention. The "attention economy" is all about grabbing as much of your attention as possible. Our challenge is to better control and direct our attention against increasingly aggressive plays for our mind-space. Meeting that challenge is only going to become more crucial as our reward-delivery technologies become even more effective.

ON THE ANXIETY
OF REDUCTION

Reading this book may lead you to a realization that feels a little uncomfortable, at least at first. I say that not in some detached way, like a moralizing narrator, but as someone who is in touch with the discomfort. I felt a twinge of it when I started writing, and I became better in touch with it as the journey continued. As I said earlier, my entire reason for even starting the journey was to attempt to answer or at least work through my questions about decision making, learning, habits, bias, and many other factors—in short, to understand *why we think as we think and do as we do*.

The uncomfortable part is the realization that so much of how we think and why we act as we do is pinned to neuro-biological dynamics. I've never been a big fan of mechanistic thinking, and I'm not interested in becoming one of its endorsers, but even the most romantic of souls must admit that there's a machine at work in all that we are. It's not a machine like your car or laptop, but one infinitely more complex, with a lot of murkiness in the machinery.

And yet, it's a kind of machine, and it runs in line with certain principles that we can grasp, with certain tendencies and machinations that we can glimpse. One such is the "reward center," which I've touched on throughout the book. When you begin to really understand what the brain's reward center is about, you get in touch with its integral role across our lives. It's not merely about seeking pleasure, feeling good, or any of the other simplistic explanations you see in popular media. It's the central pathway that enables us to learn, love, and achieve—and to become compulsive, impulsive, and addicted.

I'm especially fascinated by the reward center because it's an excellent example of the light–dark principle of our human capacities. The energy that fuels drive and passion is vital when we need it, but the same energy fuels impulsiveness and rage, which are dangerous and self-destructive. The mojo that moves us forward to learn and master knowledge is the same that can enslave us to compulsive behaviors, which lead to our unraveling. The spark that helps keep us safe by alerting us to danger is the same that can lock us in anxiety-ridden cages we feel incapable of escaping.

In all cases, you can see the "light–dark" at work. It's an inescapable reality—perhaps the most inescapable of all our realities—that we live on the slide of these spectrums, and how far we slide in either direction changes our lives.

Why should knowing this make you, or me, uncomfortable? I think the first reason is that the knowledge is at odds with what we're coached to believe from infancy onward. In a sense, we are covered with delusions layered on us by cultural forces from the moment the doctor hears the telltale cry in the delivery room. Humans, at least in Western cultures, are trained to believe that we're set apart from the rest of the animal world. Many of us are fitted with a set of blinders that force our perspectives into insulated silos radically out of touch with our inner realities.

Later on, we start running headlong into situations that challenge our blindness. The more we run into them, the more we fling and flail in ways out of line with what was "supposed" to work for us, the more questions we ask. Some of us look for answers from those whom author and philosopher Michael Novak called the "mythmakers"—those powers and personalities who exert tremendous influence in any culture. Maybe they're the same mythmakers who have been in our lives from early on, or maybe they're new ones we seek out as the vicissitudes of life drive us to find answers we haven't found through the normal channels. Others of us look elsewhere, to science and philosophy. Still others consult a mishmash of all of the above and more. Whatever the case, we're looking for answers because the delusional layering didn't deliver.

The irony of all of this searching is that *we're carrying around the answers*. What we really need is a means to understand what's resident within us. Not in a mystical sense, but in a self-diagnostic sense. We can't entirely rely on science to deliver this either, but I'd argue that scientific thinking—applied to multiple categories of knowledge—is the best tool for developing a self-diagnostic approach. This book is a compendium of self-diagnostics, using what I've termed "science-help" (rather than self-help) to achieve more clarity about why we think as we think and do as we do. Science gives us the starting points, the clues, to delve deeper; and science-help is about moving forward from those starting points.

This brings us back to the discomfort—the anxiety—that comes with this knowledge—what I call the "anxiety of reduction." As you develop more clarity around your inner reality—the symphony of dynamics that make you who and what you are—it's tempting to ruminate about just how much of "you" is a matter of predictable biology. Not predictable in a simple sense (there's nothing simple about it), but predictable in the sense that if we can know the factors and how they interact, we can generally predict a few probable outcomes, whether they include anger, fear, happiness, anxiety, guilt, joy, regret, sadness, compulsion, addiction, and so forth. If enough about someone's inner world is disclosed, then we can trace the linkages between likely cause and probable effect; we can glimpse the electrochemical dynamics playing out; and we can make reasonable assessments about what's coming next. To a certainty? No. And without notable exceptions? Of course not. But across the averages of human experience, we can get pretty close to understanding how the dots connect and what images they'll form.

That's the source of the anxiety, because once you grasp this for yourself, about yourself, you have to part with the delusion of being set apart. Far more makes us similar to each other than different from each other. Far more links us with the rest of the animal world than separates us from it. Far more defines us as biological beings than as beings hovering above our biology.

And yet, we are also creatures of the sublime—we are also more than the sum of our parts, and that's the counterbalance to the anxiety of reduction. The very fact that we're capable of the immense range of emotional expression humans embody is proof that we can both not believe we're set apart and yet still marvel at how amazing human existence truly is. The more we understand—the more clarity we can glimpse about what's going on behind the veil—the more we should want to applaud, and that's as true for our self-assessments as it is for our overall assessment of how humans work.

Having the diagnostic tools and clues to figure out why we think as we think and do as we do should ultimately *increase* our amazement, not deflate it. We part with the delusions and replace them with grounded appreciation.

Take, for example, the title of this book: *What Makes Your Brain Happy and Why You Should Do the Opposite.* It's a little tongue-in-cheek, sure. But also, you may agree at this point, it's a quite-accurate statement about the strategy we must apply to deal with our own brain, concerning so many parts of our lives. There's only one way to arrive at an understanding that enables seeing how important that strategy is for our well-being, and it's all about grounded self-diagnostics: using scientific thinking (not just a blanket application of "science" as if science has some magical powers of its own, but *scientific thinking*) to delve deeper, see the patterns, understand the triggers, trace the connections, and anticipate the outcomes.

Little by little, over time, we're getting there. Grounded self-diagnostics are made possible by the application of scientific thinking to hard problems, so expecting mountains to move without assiduous effort applied over time isn't reasonable. What is reasonable is steady progress along critical paths.

Onward.

ACKNOWLEDGMENTS

I have so many people to thank for their help in writing this book, but none more than my family members who have supported me day after day, week after week, month after month. My kids, Devin, Collin, and Kayla, keep me pushing forward on this and every project I take on. I have no greater source of energy than their love and support.

I certainly would not have had the opportunity to write this book if not for the tireless efforts of my agent, Jill Marsal, whose advice and direction have proven invaluable. I would also like to thank friends who lent their time to help me think through arguments, edit documents, and provide feedback at various points along the way, including Jeff Neale, John Vick, Robert Vandervoort, Donald Wilson Bush, and Todd Essig. I have also benefited from an incredible network of fellow writers who generously gave me their time when I needed it most. David Dobbs offered priceless advice on where to begin as a science writer trying to gain traction in the marketplace. Mark Changizi provided tips from his experience as an author and was also gracious enough to refer me to his agent (who then became my agent as well). Carl Zimmer spent time reading through my original book proposal and offered useful feedback on how to make it stronger. Robert Burton offered lessons from his experience as an author that helped guide my steps moving forward. Daniel Simons provided extremely useful input as I was preparing the final draft of this manuscript.

Special thanks goes to Wray Herbert, who generously agreed to write the foreword to this book, and well before then set a standard for how to write solid psychology exposition that served as a model for me from day one. I am also indebted to my first editor at *Scientific American Mind*, Karen Schrock, for trusting in my ability to deliver quality content to a top-tier science magazine. As well, I would like to thank Ryan Sager, one of the best all-around journalists I know, who introduced me to the crew at *True/Slant*: Coates Bateman, Andrea Spiegel, Michael Roston, and Lewis Dvorkin, most of whom I am still working with at *Forbes*, and all of whom have contributed to my career.

Early on, when I first started my blog, *Neuronarrative*, I was fortunate enough to meet several people eager to help me succeed, including a multitude of scientists, authors, and others who gave me interview time without reservation. Those relationships have been crucial to everything else I've pursued since then; I'd like to thank everyone who gave me far more than the time of day, even when I cold-called you out of the blue.

A great deal of thanks also goes to all of the talented researchers who work diligently at research institutions and labs around the world to provide new understandings of complex behavioral issues. Without their tireless efforts, none of us writing in this field would have much to talk about.

Thanks and appreciation also go to the editorial staffs at *Forbes*, *Psychology Today*, and *Scientific American Mind* who have been tremendously supportive and helpful throughout our working relationship.

NOTES

INTRODUCTION: HACKING THE COGNITIVE COMPASS

1. Simon LeVay, *The Sexual Brain* (Cambridge, MA: MIT Press, 2004).
2. John Searle, *The Rediscovery of the Mind* (Cambridge, MA: MIT Press, 1992).
3. Greg Miller, "Growing Pains for fMRI," *Science* (June 2008): 1412–14.
4. Todd Essig, PhD, in discussion with the author, March 2011.
5. Daniel Kahneman and Amos Tversky, "On the Reality of Cognitive Illusions," *Psychological Review* 103 (July 1996): 582–91.

CHAPTER 1: ADVENTURES IN CERTAINTY

1. Pamela S. Turner, "Showdown at Sea: What Happens When Great White Sharks Go Fin-to-Fin with Killer Whales?" *National Wildlife Federation*, October 1, 2004, http://www.nwf.org/News-and-Magazines/National-Wildlife/Animals/Archives/2004/Showdown-at-Sea.aspx (accessed June 25, 2010).
2. *Telegraph*, "Killer Whales Attack and Eat Sharks," *Telegraph*, November 27, 2009, http://www.telegraph.co.uk/earth/wildlife/6668575/Killer-whales-attack-and-eat-sharks.html (accessed June 25, 2010).
3. Susan Blackmore, *The Meme Machine* (New York: Oxford University Press, 2001).
4. Ming Hsu et al., "Neural Systems Responding to Degrees of Uncertainty in Human Decision-Making," *Science* 310 (December 2005): 1680–83.
5. Jonas T. Kaplan et al., "Neural Correlates of Maintaining One's Political Beliefs in the Face of Counterevidence," *Scientific Reports* 6 (2016): 39589.
6. Robert Burton, preface to *On Being Certain: Believing We You Are Right Even When You're Not* (New York: St. Martin's Press, 2008).
7. Charles W. Eriksen, "The Flankers Task and Response Competition: A Useful Tool for Investigating a Variety of Cognitive Problems," *Visual Cognition* 2 (1995): 101–18.

8. Daniel J. Simons and Christopher F. Chabris, "Gorillas in Our Midst: Sustained Inattentional Blindness for Dynamic Events," *Perception* 28 (June 1999): 1059–74.

9. Amos Tversky and Daniel Kahneman, "The Framing of Decisions and the Psychology of Choice," *Science* 12 (June 1981): 453–58.

10. Sam Keen told this story during a workshop titled "Your Mythic Journey," which was recorded on audio tape for distribution by Shambhala Press (1996).

11. Keith E. Stanovich and Richard F. West, "Individual Differences in Reasoning: Implications for the Rationality Debate?" *Behavioral and Brain Sciences* 23 (2000): 645–726.

12. Frank Miller, *300* (Milwaukie, OR: Dark Horse, 1999).

13. Rossella Lorenzi, "Shroud of Turin Secretly Hidden," *Discovery News*, April 6, 2009, http://dsc.discovery.com/news/2009/04/06/turin-shroud-templars.html (accessed July 5, 2010).

14. Fox News, "Does Hidden Text Prove Shroud of Turin Real?" Fox News, November 20, 2009, http://www.foxnews.com/scitech/2009/11/20/does-hidden-text-prove-shroud-turin-real/ (accessed July 5, 2010).

15. Archie O. de Berket et al., "Computations of Uncertainty Mediate Acute Stress Responses in Humans," *Nature Communications* 7 (2016); published online March 29, 2016 (Berkeley: University of California Press, 2007), p. 214.

16. "Kiai Master vs MMA," YouTube video, 3:35, posted by "vortexblade," January 3, 2007, http://www.youtube.com/watch?v=gEDaCIDvj6I (accessed July 10, 2010).

17. Robert Axelrod, "Schema Theory: An Information Processing Model of Perception and Cognition," *American Political Science Review* 67 (December 1973): 1248–66.

18. Walter Bradford Cannon, *Wisdom of the Body* (New York: W. W. Norton, 1963).

19. Sam Harris et al., "The Neural Correlates of Religious and Nonreligious Belief," *PLoS ONE* 4 (2009), http://www.plosone.org/article/infopercent3Adoi percent2F10.1371 percent2Fjournal.pone.0007272 (accessed May 15, 2011).

20. Bruce M. Hood, *SuperSense: Why We Believe in the Unbelievable* (New York: HarperCollins, 2009).

21. A. W. Kruglanski, J. Y. Shah, A. Pierro, and L. Mannetti, "When Similarity Breeds Content: Need for Closure and the Allure of Homogeneous and Self-Resembling Groups," *Journal of Personality and Social Psychology* 83, no. 3 (September 2002): 648–62.

CHAPTER 2: SEDUCTIVE PATTERNS AND SMOKING MONKEYS

1. Ernest Gallo, "Jung and the Paranormal," in *The Encyclopedia of the Paranormal*, ed. Gordon Stein (Amherst, NY: Prometheus Books, 1996).

2. Story related to the author by a confidential source, March 2011.

3. Stephen Grossberg, "The Link between Brain Learning, Attention, and Consciousness," *Consciousness and Cognition* 8, no. 1 (March 1999): 3–45.

4. Daniel Dennett et al., "The Intentional Stance in Theory and Practice," in *Machiavellian Intelligence: Social Expertise and the Evolution of Intellect in Monkeys, Apes, and Humans*, ed. Richard W. Byrne and Andrew Whiten (New York: Clarendon/Oxford University Press), pp. 180–202.

5. Paul Rogers et al., "Paranormal Belief and Susceptibility to the Conjunction Fallacy," *Applied Cognitive Psychology* 23 (June 2008): 524–42.

6. David A. Moscovitch and Kristin Laurin, "Randomness, Attributions of Arousal, and Belief in God," *Psychological Science* 21 (February 2010): 216–18.

7. An excellent overview of probability and its outcomes can be found in Leonard Mlodinow, *The Drunkard's Walk: How Randomness Rules Our Lives* (New York: Random House, 2009).

8. Helena Matute et al., "Illusion of Control in Internet Users and College Students," *CyberPsychology & Behavior* 10 (April 2007): 176–81.

9. Kyle Siler, "Social and Psychological Challenges of Poker," *Journal of Gambling Studies* 26 (December 2009): 401–20.

CHAPTER 3: WHY A HAPPY BRAIN DISCOUNTS THE FUTURE

1. D. Read and B. van Leeuwen, "Predicting Hunger: The Effects of Appetite and Delay on Choice," *Organizational Behavior and Human Decision Possesses* 76 (1998): 189–205.

2. Dasgupta Partha and Eric Maskin, "Uncertainty and Hyperbolic Discounting," *American Economic Review* 95 (2005): 1290–99.

3. *Primetime: What Would You Do?* is a production of ABC hosted by anchor John Quiñones.

4. Roger Beuhler and Kathy McFarland, "Intensity Bias in Affective Forecasting:

The Role of Temporal Focus," *Personality and Social Psychology Bulletin* 27 (2001): 1480–93.

CHAPTER 4: THE MAGNETISM OF AUTOPILOT

1. Kalina Christoff et al, "Mind-Wandering as Spontaneous Thought: A Dynamic Framework," *Nature Reviews Neuroscience* 17 (September 2016): 718–31.

2. Jerome Singer was also among the first researchers to draw a link between mind wandering and creativity; his work remains influential in current studies probing this topic.

3. Randy Buckner et al., "The Brain's Default Network: Anatomy, Function, and Relevance to Disease," *Annals of the New York Academy of Sciences* 1124 (2008): 1–38.

4. John Cleese spoke at the Creativity World Forum in Belgium on the topic of how to become more creative; his speech, titled "De Bron von Creativiteit," was recorded and subsequently released on YouTube: "John Cleese WCF," YouTube video, 10:37, posted by "zeekomkommers," March 18, 2009, http:// www.youtube .com/ watch?v=zGt3-fxOvug (accessed September 14, 2011).

5. M. A. Killingsworth and Daniel T. Gilbert, "A Wandering Mind Is an Unhappy Mind," *Science* 12 (November 2010): 932.

6. Edward R. Watkins, "Constructive and Unconstructive Repetitive Thought," *Psychological Bulletin* 134 (March 2008): 163–206.

7. Bertrand Russell, *The Conquest of Happiness* (London: Liveright, 1996), p. 65.

8. C. K. Hesse et al., "Idleness Aversion and the Need for Justifiable Busyness," *Psychological Science* 21 (July 2010): 926–30.

9. Kalina Christoff et al., "Experience Sampling during fMRI Reveals Default Network and Executive System Contributions to Mind Wandering," *Proceedings of the National Academy of Sciences* 26 (May 2009): 8719–24.

10. Michael A. Sayette et al., "Lost in the Sauce? The Effects of Alcohol on Mind Wandering," *Psychological Science* 20 (June 2009): 747–52.

11. Robert Spunt et al., "The Default Mode of Human Brain Function Primes the Intentional Stance," *Journal of Cognitive Neuroscience* 27 (June 2015): 1116–24.

12. Jeff Moher et al., "Dissociable Effects of Salience on Attention and Goal-Directed Action," *Current Biology* 26 (July 2016): 2040–46.

13. Quoted from an article I wrote for the online edition of *Forbes* magazine, "When It Comes to Dealing with Distractions, Sweating the Small Stuff Makes Scientific Sense," July 28, 2015, https://www.forbes.com/sites/

daviddisalvo/2015/07/28/when-it-comes-to-dealing-with-distractions-sweating-the-small-stuff-makes-scientific-sense/#536b3e4f2f9f (accessed October 19, 2017).

14. Ibid.

15. Cary Stothart et al., "The Attentional Cost of Receiving a Cell Phone Notification," *Journal of Experimental Psychology: Human Perception and Performance* 41 (August 2015): 893–97.

CHAPTER 5: IMMERSION AND THE GREAT ESCAPE

1. *Multiplicity* by Rita Carter provides an excellent overview of multiple-selves theory and its implications. Rita Carter, *Multiplicity: The New Science of Personality, Identity, and the Self* (Boston: Little, Brown, 2008).

2. James McCurry, "Internet Addiction Driving South Koreans into Realms of Fantasy," *Guardian*, July 2010, http://www.guardian.co.uk/ world/2010/jul/13/internet-addiction-south-korea (accessed July 2010).

3. Todd Essig, PhD, in discussion with the author, March 2011.

4. Ibid.

5. George F. Koob et al., "Neurocircuitry of Addiction," *Neuropsychopharmacology* 35 (January 2010): 217–38.

6. W. M. Doyan et al., "Dopamine Activity in the Nucleus Accumbens during Consummatory Phases of Oral Ethanol Self-Administration," *Alcoholism, Clinical and Experimental Research* 10 (October 2003): 1573–82.

7. Gary Small, MD, in discussion with the author, June 2010.

8. Gary Null, *Get Healthy Now* (New York: Seven Stories, 2006), p. 269.

9. Scott Caplan, PhD, in discussion with the author, February 2010.

10. Scott Caplan et al., "Relations among Loneliness, Social Anxiety, and Problematic Internet Use," *CyberPsychology and Behavior* 10 (April 2007): 234–42.

11. Caplan, discussion with the author.

12. Jay L. Derrick et al., "Social Surrogacy: How Favored Television Programs Provide the Experience of Belonging," *Journal of Experimental Social Psychology* 45 (February 2009): 352–62.

13. John Cacioppo et al., "Loneliness as a Specific Risk Factor for Depressive Symptoms: Cross-Sectional and Longitudinal Analyses," *Psychology and Aging* 21 (March 206) 140–51.

14. Anna Abraham et al., "Reality = Relevance? Insights from Spontaneous Modulations of the Brain's Default Network When Telling Apart Reality from

Fiction," *PLoS ONE* 4 (2009), http://www.plosone.org/article/ infopercent3Adoi percent2F10.1371 percent2Fjournal.pone.0004741 (accessed May 15, 2011).

15. John Cacioppo et al., "Alone in the Crowd: The Structure and Spread of Loneliness in a Large Social Network," *Journal of Personality and Social Psychology* 97 (December 2009): 977–91.

CHAPTER 6: REVVING YOUR ENGINE IN IDLE

1. William Hart, "The Effects of Chronic Achievement Motivation and Achievement Primes on the Activation of Achievement and Fun Goals," *Psychology Bulletin* 135 (July 2009): 555–88.

2. Stephen Garcia, "The N-Effect: More Competitors, Less Competition," *Psychological Science* (July 2009): 871–77.

3. Keri Kettle et al., "Motivation by Anticipation: Expecting Rapid Feedback Enhances Performance," *Psychological Science* (April 2010): 545–47.

4. Peter Sokol-Hessner et al., "Thinking Like a Trader Selectively Reduces Individuals' Loss Aversion," *Proceedings of the National Academy of Sciences* 106 (July 2008): 5035–40.

5. Prashanth U. Nyer et al., "Public Commitment as a Motivator for Weight Loss," *Psychology and Marketing* 27 (January 2010): 1–12.

6. L. Senay et al., "Motivating Goal-Directed Behavior through Introspective Self-Talk: The Role of the Interrogative Form of Simple Future Tense," *Psychological Science* 21 (April 2010): 499–504.

CHAPTER 7: WRITING PROMISES ON AN ETCH-A-SKETCH

1. L. F. Nordgren et al., "The Restraint Bias: How the Illusion of Self-Restraint Promotes Impulsive Behavior," *Psychological Science* 20 (December 2009): 523–28.

2. Theodore Evans et al., "Chimpanzees Use Self-Distraction to Cope with Impulsivity," *Biological Letters* 3 (August 2007): 599–602.

3. D. M. Wegner et al., "Transactive Memory in Close Relationships," *Journal of Personality and Social Psychology* 61 (December 1991): 923–29.

4. Gráinne M. Fitzsimons et al., "Outsourcing Self-Regulation," *Psychological Science* 22 (2011): 369–75.

5. Carey K. Morewedge et al., "Thought for Food: Imagined Consumption Reduces Actual Consumption," *Science* 10 (December 2010): 1530–33.

6. Janet Polivy et al., "Getting a Bigger Slice of the Pie: Effects on Eating and Emotion in Restrained and Unrestrained Eaters," *Appetite* 55 (December 2010): 426–30.

7. Eva Pool et al., "Stress Increases Cue-Triggered 'Wanting' for Sweet Reward in Humans," *Journal of Experimental Psychology: Animal Learning and Cognition* 41, no. 2 (April 2015): 128–36.

8. S. Sachdeva et al., "Sinning Saints and Saintly Sinners: The Paradox of Moral Self-Regulation," *Psychological Science* 20 (2009): 523–28.

CHAPTER 8: WANT, GET, REGRET, REPEAT

1. At this writing, there are no fewer than 28,000 scholarly entries for research related to eBay auctions available via Google Scholar and 2,700 via ScienceDirect.com.

2. Michael Wiederman, "Why It's So Hard to Be Happy," *Scientific American Mind* (February/March 2007).

3. Ibid.

4. Ab Lit et al., "Lusting while Loathing: Parallel Counterdriving of Wanting and Liking," *Psychological Science* 21 (February 2010): 118–25.

5. Kai Epstude et al., "The Functional Theory of Counterfactual Thinking," *Personality and Social Psychology Review* 12 (May 2008): 168–92.

6. A. C. Quelhas et al., "Counterfactual Thinking and Functional Differences in Depression," *Clinical Psychology and Psychotherapy* 15 (September 2008): 352–65.

CHAPTER 9: SOCIALIZING WITH MONKEYS LIKE US

1. Laurie Santos, PhD, in discussion with the author, January 2011.

2. Daniela Schiller et al., "A Neural Mechanism of First Impressions," *Nature Neuroscience* 12 (2009): 508–14.

3. Paul Zak, "How to Run a Con," *Psychology Today* (blog), November 2008, http://www.psychologytoday.com/blog/the-moral-molecule/200811/ how-run-con (accessed November 22, 2008).

4. Bryan Gibson et al., "How the Adoption of Impression Management Goals

Alters Impression Formation," *Personality and Social Psychology Bulletin* 36 (October 2010): 1543–54.

5. Gerald Mollenhorst et al., "Social Contexts and Personal Relationships: The Effect of Meeting Opportunities on Similarity for Relationships of Different Strength," *Social Networks* 30 (January 2008): 60–68.

6. Hashimoto Yamagishi et al., "Preferences versus Strategies as Explanations for Culture-Specific Behavior," *Psychological Science* 19 (2008): 579–84.

7. Elizabeth Tricomi et al., "Neural Evidence for Inequality-Averse Social Preferences," *Nature* 463 (February 2010): 1089–91.

8. Xiaoming Jiang et al., "On How the Brain Decodes Vocal Cues about Speaker Confidence," *Cortex* (May 2015): 9–34.

9. Toshio Yamagishi et al., "The Private Rejection of Unfair Offers and Emotional Commitment," *Proceedings of the National Academy of Sciences* 106 (July 2009): 11520–23.

CHAPTER 10: THE GREAT TRUTH RUB-OFF

1. Jan B. Engelmann et al., "Expert Financial Advice Neurobiologically 'Offloads' Financial Decision Making under Risk," *PLoS ONE* 4 (2009), http://www.plosone.org/article/info%3Adoi%2F10.1371%2Fjournal.pone.0004957 (accessed May 20, 2010).

2. Jamil Zeki et al., "Social Influence Modulates the Neural Computation of Value" *Psychological Science* 22, no. 7 (July 2011): 894–900.

3. Christian Unkelbach, "The Learned Interpretation of Cognitive Fluency," *Psychological Science* 17 (2006): 339–45.

4. Christian Unkelbach, "Reversing the Truth Effect: Learning the Interpretation of Processing Fluency in Judgments of Truth," *Journal of Experimental Psychology: Learning, Memory, and Cognition* 33 (January 2007): 219–30.

5. Lauren Movius et al., "Motivating Television Viewers to Become Organ Donors," *Cases in Public Health Communication & Marketing*, June 2007, http://www.gwumc.edu/sphhs/departments/pch/phcm/casesjournal/ volume1/peer-reviewed/cases_1_08.pdf (accessed July 18, 2011).

6. Paul Thibodeau et al., "Metaphors We Think With: The Role of Metaphor in Reasoning," *PLoS ONE* (2011), http://dx.doi.org/10.1371/ journal.pone.0016782 (accessed July 15, 2011).

7. William Shakespeare, *Romeo and Juliet*, II.ii.47–48.

8. Lera Boroditsky, "How Language Shapes Thought," *Scientific American* 304 (January 2011): 62–65.

9. Daniel Casasanto et al., "When Left Is 'Right': Motor Fluency Shapes Abstract Concepts," *Psychological Science* 22 (2011): 419–22.

CHAPTER 11: HOW YOUR BRAIN CATCHES PSYCHOSOCIAL COLDS

1. James H. Fowler et al., "The Dynamic Spread of Happiness in a Large Social Network," *British Medical Journal* 337 (December 2008).

2. K. Haegler et al., "No Fear, No Risk! Human Risk Behavior Is Affected by Chemosensory Anxiety Signals," *Neuropsychologica* 48 (September 2010): 3901–3908.

3. James H. Fowler et al., "The Spread of Obesity through a Large Social Network over 32 Years," *New England Journal of Medicine* 357 (July 2007): 370–79.

4. Carrie Arnold, "We're in This Together," *Scientific American Mind* (May 2011).

5. Nathanael J. Fast et al., "Blame Contagion: The Automatic Transmission of Self-Serving Attributions," *Journal of Experimental Social Psychology* 46 (January 2010): 97–106.

6. Joshua M. Ackerman et al., "You Wear Me Out: The Vicarious Depletion of Self-Control," *Psychological Science* 20 (March 2009): 326–32.

7. Matthew Campbell et al., "Ingroup–Outgroup Bias in Contagious Yawning by Chimpanzees Supports Link to Empathy," *PLoS ONE* 6 (2011), http://www.plosone.org/article/infopercent3Adoipercent2F10.1371percent2Fjournal.pone.0018283 (accessed June 14, 2011).

8. M. Stel et al., "You Want to Know the Truth? Then Don't Mimic!" *Psychological Science* 20 (April 2009): 693–99.

9. Lane Beckes et al., "Familiarity Promotes the Blurring of Self and Other in the Neural Representation of Threat," *Social Cognitive and Affective Neuroscience* 8, no. 6 (2013): 670–77.

10. Material in this text box is pulled from David DiSalvo, "Study: To the Human Brain, Me Is We," *Forbes*, August 22, 2013, https://www.forbes.com/sites/daviddisalvo/2013/08/22/study-to-the-human-brain-me-is-we/#131051eff1a0 (accessed October 26, 2017).

11. Paul Bloom, *Against Empathy: The Case for Rational Compassion* (New York: Ecco, an imprint of HarperCollins, 2016).

CHAPTER 12: THE HIDDEN POWER OF STUFF

1. Nils B. Jostman et al., "Weight as an Embodiment of Importance," *Psychological Science* 20 (February 2009): 1169–74.

2. Kendal Eskine et al., "A Bad Taste in the Mouth: Gustatory Disgust Influences Moral Judgment," *Psychological Science* 22 (November 2010): 295–99.

3. H. IJzerman et al., "The Thermometer of Social Relations: Mapping Social Proximity on Temperature," *Psychological Science* 20 (October 2009): 1214–20.

4. Joshua M. Ackerman et al., "Incidental Haptic Sensations Influence Social Judgments and Decisions," *Science* 25 (June 2010): 1712–15.

5. J. Witt, "How the Body Shapes the Mind," (paper presented at the meeting of the American Association for the Advancement of Science, Washington, DC, February 2016).

6. Jessica Witt et al., "Perceived Distance and Obesity: It's What You Weigh, Not What You Think," *Acta Psychologica* 165 (March 2016):1–8.

CHAPTER 13: YOUR MIND IN REWRITES

1. The material in this section comes from multiple sources, including Pierce J. Howard, *The Owner's Manual for the Brain* (Austin: Bard, 2006), pp. 93–109; and Jeanette Norden, *Understanding the Brain: Course Guidebook* (Chantilly, VA: Great Courses), pp. 108–12.

2. Adam Blake et al., "The Apple of the Mind's Eye: Everyday Attention, Metamemory, and Reconstructive Memory for the Apple Logo," *Quarterly Journal of Experimental Psychology* (February 2015): 858–65.

3. Linda Henkel et al., "Photograph-Induced Memory Errors: When Photographs Make People Claim They Have Done Things They Have Not," *Applied Cognitive Psychology* 25 (2011): 78–86.

4. Robert Nash et al., "Digitally Manipulating Memory: Effects of Doctored Videos and Imagination in Distorting Beliefs and Memories," *Memory and Cognition* 37 (2009): 414–24.

5. B. Zhu et al., "Treat and Trick: A New Way to Increase False Memory," *Applied Cognitive Psychology* 24 (2009): 1199–208.

6. C. van Golde et al., "High-Prevalence Information from Different Sources Affects the Development of False Beliefs," *Applied Cognitive Psychology* 24 (2009): 152–63.

7. Alex Haynes et al., "A Surgical Safety Checklist to Reduce Morbidity and Mortality in a Global Population," *New England Journal of Medicine* 360 (January 2009): 491–99.

8. Cristina M. Atance et al., "Episodic Future Thinking," *Trends in Cognitive Science* 5 (December 2001): 533–39.

CHAPTER 14: BORN TO COPY, LEARN TO PRACTICE

1. Derrick Lyons et al., "The Hidden Structure of Overimitation," *Proceedings of the National Academy of Sciences* 104, no. 50 (December 11, 2007): 19751–56.

2. K. Anders Ericsson, "Deliberate Practice and Acquisition of Expert Performance: A General Overview," *Academic Emergency Medicine* 15 (November 2008): 988–94.

3. K. Anders Ericsson, *The Cambridge Handbook of Expertise and Expert Performance* (Cambridge: Cambridge University Press, 2006), pp. 685–703.

4. David DiSalvo, "Week in Ideas," *Wall Street Journal*, March 26, 2011.

5. Merim Bilalic et al., "Specialization Effect and Its Influence on Memory and Problem Solving in Expert Chess Players," *Cognitive Science* 33, no. 6 (August 2009): 1118–43.

6. Christian Jarrett, "The Expert Mind of a Burglar," *BPS Research Digest Blog*, September 2006, http://bps-research-digest.blogspot.com/2006/09/ expert-mind-of-burglar.html (accessed May 15, 2011).

CHAPTER 15: MIND THE GAP

1. Yoshahisha Kashima et al., "The Self-Serving Bias in Attributions as a Coping Strategy," *Journal of Cross-Cultural Psychology* 17 (March 1986): 83–97.

2. Jeremy Dean, "The Zeigarnik Effect," *Psyblog*, http://www.spring.org. uk/2011/02/the-zeigarnik-effect.php (accessed May 15, 2011).

3. Bastien Blain et al., "Neural Mechanisms Underlying the Impact of Daylong Cognitive Work on Economic Decisions," *Proceedings of the National Academy of Sciences* 113, no. 25 (October 2015): 6967–72.

4. Paul Ekman Group, "Dr. Paul Ekman," http://www.paulekman.com/about-ekman/ (accessed May 15, 2011).

5. Sam Harris et al., "The Neural Correlates of Religious and Non-Religious Belief," *PLoS ONE* 4 (2009), http://www.plosone.org/article/infopercent3Adoi percent2F10.1371 percent2Fjournal.pone.0007272 (accessed May 15, 2011).

6. Elizabeth B. Raposa et al., "Prosocial Behavior Mitigates the Negative Effects of Stress in Everyday Life," *Clinical Psychological Science* 4 (December 2015): 691–98.

7. Bernd Weber et al., "The Medial Prefrontal Cortex Exhibits Money Illusion," *Proceedings of the National Academy of Sciences* 106 (March 2009): 5025–28.

8. Paul Zak, "How to Run a Con," *Psychology Today* (blog), November 2008, http://www.psychologytoday.com/blog/the-moral-molecule/200811/how-run-con (accessed May 15, 2011).

9. Lera Boroditsky, "How Language Shapes Thought," *Scientific American* 304 (January 2011): 62–65.

SPECIAL SECTION 2: OF TECHNOLOGY AND REWARDS

1. Robert Sapolsky, *Behave: The Biology of Humans at Our Best and Our Worst* (London: Penguin, 2017), pp. 64–77.

2. W. Schultz, "Dopamine Signals for Reward Value and Risk: Basic and Recent Data," *Behavioral and Brain Functions* 6 (2010): 24.

3. Robert Sapolsky, "The Dopamine Jackpot! Sapolsky on the Science of Pleasure," speech given at the California Academy of Sciences, San Francisco, California, March 2, 2011, video, 1:16:08, http://library.fora.tv/2011/02/15/Robert _Sapolsky_Are_Humans_Just_Another_Primate (accessed October 27, 2017).

4. Sharon Begley, *Can't Just Stop: An Investigation of Compulsions* (New York: Simon and Schuster, 2017), pp. 121–37.

5. T. Wilson et al., "Just Think: The Challenges of the Disengaged Mind," *Science* 345 (July 2014): 75–77.

6. A. K. Przybylskia et al., "Motivational, Emotional, and Behavioral Correlates of Fear of Missing Out," *Computers in Human Behavior* 29 (July 2013): 1841–48.

INDEX

cortex; dorsomedial prefrontal cortex; medial prefrontal cortex; parietal cortex; posterior cingulate cortex; visual cortex

counterfactual thinking, 139, 140, 238–39

creativity
creative problem solvers not needing as much closure, 54–55
link between mind wandering and creativity, 81–82

credibility-based positions, 47–48

CSI (TV show), 165

culture
Chinese street vendors, 40–41
influences on, 286
modern cultures and brain capabilities, 61, 279
moving faster than natural selection, 145–46
orca culture, 30–31

Cure: A Journey into the Science of Mind Over Body (Marchant), 266

Damasio, Antonio, 259

D&D game, 90–91

danger, skydiving as an example of brain handling, 53–54

dating and setting up bodily perceptions, 188

daydreaming, 80, 81, 85. *See also* autopilot, being on

deception and con artists, 148

deceptive interactions and mimicry, 180

decision making, 258
automatic decision making, 11
beating up self over poor decisions, 139–40
framing of a decision maker, 38
"offloading" decisions to external sources, 158–61
purchasing decisions based on group behavior, 161–62
regret as a factor in, 136–38
weight affecting process of, 184–85

declarative memory, 195

default-mode processing, 85

"default network," 80–81, 83. *See also* medial prefrontal cortex; parietal cortex; posterior cingulate cortex

defensiveness, 79, 155, 266

delaying gratification. *See* gratification

deliberate practice (practicing for a purpose), 214–17

deliberation, 9, 11, 38

Dellande, Stephanie, 115

delusion of being set apart, 286–87

demonstration, use of in learning, 30

Dennett, Daniel, 64, 259

depression, 250
and counterfactual thinking, 139, 239
and obsessive rumination, 83–84

Descartes, René, 21

Dewey, John, 193

Dialogues Concerning Natural Religion (Hume), 17

DiSalvo, David, 9–10, 11, 12, 13

rumination, 83–84. *See also* thinking process
Russell, Bertrand, 82, 269
Ryukerin (Kiai master), 49–50

Safe Surgery Saves Lives Program (World Health Organization), 206
Santos, Laurie, 145–46, 153
Sapolsky, Robert M., 269
satisfaction and stress, 129
schemata, 51–52, 163, 241
Schwartz, Barry, 270
Science Friction: Where the Known Meets the Unknown (Shermer), 270
science-help (versus self-help), 13, 23, 222, 287
Scientific American Brave New Brain, The (Horstman), 263
scientific thinking, 287, 288
scripts, 9–10, 11, 39
Secret, The (Byrne), 60
selective attention (selectivity bias), 33–36, 53
    memory selectivity, 197–98
        visual memory, 199
    See also availability bias; framing bias
self, sense of. *See* sense of self
self-awareness, 111
Self Comes to Mind: Constructing the Conscious Brain (Damasio), 259
self-control, 119–31
    chimpanzee showing, 123
    and goals, 120, 124, 127
    illusion of control, 67–68

impact of the "oh, what the hell" effect, 126–28
and the "moral self-regulation effect," 130–31, 241–42
observing others exert self-control, 177–79
outsourcing to others (transactive self-control), 124–25
and restraint bias, 121–24, 231
transactive self-control, 124
use of imagination to reduce temptations, 125–26, 231
See also impulse control
self-diagnostics, 287, 288
self-evaluation, 243
self-help advice, 13, 22–23, 24, 59, 124, 222, 287
self-image protection, 176
self-motivation, 117–18, 218, 229
self-narrative, 253
self-negotiation
self-perception, 122, 124, 224
self-regulation. *See* balance and moral self-regulation
self-restraint, 120, 121, 125–26
self-serving bias, 227
self-talk, 117, 118, 229
selling a message, 162–64
semantic memory, 195
sense of self
    blurring of in echopraxia, 210
    challenging sense of self, 158
    "default network" integral to, 80
    impact of amPFC and PCC on, 99